Smartlipo™

Melting Away Fat

Presented By

Jeffry B. Schafer, MD FRSM

Published by MWP Media
ISBN 13: 978-0-9818209-2-7
ISBN 10: 0-9818209-2-1

Cover: Art design by Han Le.

Disclaimer

The information presented in this book is not intended to replace medical advice of your personal physician or any surgeon performing Smartlipo or Liposuction. In addition, those associated with this book recommend a proper doctor-patient relationship be established for patients seeking Smartlipo.

Dedication

This book is dedicated to my loving wife, Sandy, and our standard poodle daughter, Joelle.

Acknowledgments

I would like to thank the able surgical and administrative staff of New Image Cosmetic Surgery: Joe Kraut, Physician Assistant, Buckley Lewis, Medical Assistant, Laine Alves, LVN, Elizabeth Lafleur, Medical Assistant, Tammy Ladderbush, Consultant, Eliza Arevalo, Receptionist and Shiela Sapitan, Medical Assistant.

Table of Contents

Chapter **Page**

1 *The Latest in Laser Technology* 1

2 *Stubborn Fat in Certain Areas* 13

3 *Overview of Results with Smartlipo* 33

4 *Laser Surgery and Recovery* 53

5 *Other Methods of Fat Removal* 67

6 *Micro Liposuction* 73

7 *Questions and Answers* 77

Glossary 83

Photos of Sculpted Patients 89

About the Author 97

The Amazing Smartlipo

The latest procedure for melting away fat is called Smartlipo. This process eliminates localized fat deposits by means of a special laser fiber housed inside a tiny tube called a cannula.

It is a more refined method of fat removal than traditional liposuction and is generally recommended for smaller areas of the body. Smartlipo is also gaining popularity for melting facial fat by re-defining the jaw line and getting rid of the discouraging double chin.

There are several advantages to Smartlipo. It is gentler and safer. This is, in part, due to the much smaller cannulas. A cannula is a hollow metal tube, usually made out of titanium or stainless steel, and is less than 20 cm long and 6 mm in width. These smaller cannulas are called micro-cannulae with its optic-fiber laser tip.

The use of a tiny cannula means smaller incisions, so there is less scarring. Additionally, the heat generated by this method of a laser tip is of a greater benefit for the patient. With the body's built-in healing response, the laser causes tissues to contract. Smartlipo is effective as a fat removal procedure with less bruising and swelling and a faster recovery. This is a breakthrough in the field of surgical fat removal.

1

Smartlipo is sweeping the country with its safe, near painless sculpting innovations that create astounding results.

The Latest in Laser Technology

Increasing numbers of people seek a solution to removing unwanted fat. Now there is a marvelous new solution to body sculpting that is sweeping the world of innovative fat removal. The new method that offers outstanding results is called Smartlipo. This procedure uses the latest in laser technology which literally melts unwanted fat in the body.

The new method allows you to have the sculpted body you've always wanted. Liposuction is the most popular surgical procedure performed in the United States for both women and men.

Smartlipo is the slightly invasive method of fat removal and has brought about a new phase of liposuction. The procedure usually lasts less than an hour. It is a body shaping procedure that can remove fat from almost anywhere of the body and is the first laser-assisted lipolysis system to offer a revolutionary and minimally invasive procedure for body toning. Using a high-powered laser, Smartlipo offers this new laser body sculpting with less down time and minimal side effects, unlike traditional liposuction.

There is no question that advanced body sculpting with Smartlipo is revolutionary in laser fat removal. Approved in late 2006 by the FDA, it is becoming one of the most sought after methods for slimming the body. Smartlipo has been effectively performed by doctors who have done traditional liposuction for two decades. They are finding remarkable results with the laser sculpting that offers the advantage of smooth results and tighter skin.

Not only is Smartlipo effective in reshaping areas of the body and face, but it is less invasive than classic liposuction and delivers more immediate results.

Smartlipo is the first laser that vaporizes fat by using a tiny laser fiber to dissolve the fat without the need of removing chunks of fat by suction. The procedure is bloodless and offers skin retraction and tightening. The tightening alone is a remarkable advance. The fat removal method of melting it away is usually performed under local anesthesia but varies with the surgeon.

A small cannula — a thin tube with a laser fiber on the tip — is inserted through a tiny puncture hole in the skin that requires no stitches for closure. In this manner there are no unsightly scars. It is an effective treatment for shaping and sculpting fat bulges, such as tummy, love handles, thighs, bra bulges, double chin, upper-arms and the hips and buttocks.

Smartlipo is described by most doctors as *very* effective compared to older, classic liposuction methods. Some patients who have struggled with stubborn fat deposits that were impossible to shed have found the new technology extremely rewarding. Classic liposuction is done by sucking out fat with broad strokes for large amounts of fat removal. This new method of Smartlipo liquefies the fat cells in a gentle way. For large amounts of fat, some surgeons combine a traditional liposuction procedure that requires suction fat removal, and then switch to the Smartlipo laser fat removal method for final sculpting. However, patients seeking to remove areas of stubborn fat, who otherwise are not burdened with excessive weight, find Smartlipo to be the answer to a finely sculpted body.

Judy is a good example of a patient who had unsightly body shape with bulging hips. As an aerobics instructor she felt a professional need to rid her hips of the slightly bulging fat that refused to leave with a rigorous daily workout.

Although the fat deposits on her hips were minor and not especially noticeable, when she stood in the gym in tights with her class of students, she felt it was obvious that she had slight bulges in her hips. It made her self-conscious to go through her routine knowing she had done everything possible with exercise to rid those slight pockets of fat. They were exercise resistant.

Judy had never considered full, traditional liposuction since she really had little excess body fat. When she heard about Smartlipo, she made an appointment for a consultation and was convinced that it was the solution to her fleshy hips. She had the surgery and it was performed in forty-five minutes. She was back in front of her aerobic class the following week, looking great. As far as her class members were concerned, it looked as if Judy had done some dynamic body exercises to get rid of the slight bulges.

With almost no evident scars where the tiny incisions were made in her hips to insert the laser-tipped cannula, she was elated with the outcome, particularly with the additional bonus that came with skin tightening in the region of the fat removal. The laser, with its infrared light source tightened her skin by heating just under the skin's surface, causing the skin to contract (tighten). It was noticeable to Judy the following days after surgery. This offered an even more sculpted appearance that continues to improve for up to 6 months.

The outpatient procedure often takes less than an hour to perform while the patient remains comfortable. In short, Smartlipo is less invasive, helping to sculpt areas of the body that diet and exercise fail to address, while skin tightens.

People that are healthy and not obese, but are simply overweight or have fat deposits to remove with the laser are the best candidates for this type of procedure. Smartlipo is not intended to significantly reduce gross body fat as is the case with most patients seeking traditional liposuction. However, as mentioned earlier, the traditional liposuction and Smartlipo are at times combined in a more extensive fat removal procedure.

An elastic bodysuit or a compression garment may have to be worn after the surgery in order to control and reduce swelling. This compression garment also helps the body mold into its new shape. Patients often see an immediate improvement, certainly within a few days following the procedure.

After the Smartlipo procedure — that some refer to as *Lunchtime Smartlipo* — most patients usually experience very few side effects. The surgery, due to its less invasive nature, leaves behind little damage to the surrounding tissue when compared to traditional liposuction.

Lasers and the Smartlipo Laser

Various methods of laser technology have been widely practiced for several decades now and are routinely applied in eye surgery, dental work and skin rejuvenation, as well as other areas of medicine. An example of an effective use of lasers is the very popular vision improvement procedure called LASIK. It has been effectively performed on millions of patients seeking improved.

The word LASER comes from the acronym *Light Amplification by Stimulated Emission of Radiation*. Lasers are precise and safe; today's medical-grade lasers work by creating a powerful light beam that is adjusted to a specific frequency range. Smartlipo employs a carefully calibrated laser to liquefy fat deposits through the skin's top layers.

Approved by the FDA, the instrument utilizes a tiny filament which delivers a series of rapid fire pulses that researchers have found helps shrink the undersurface of the skin and selectively ruptures the fat cells.

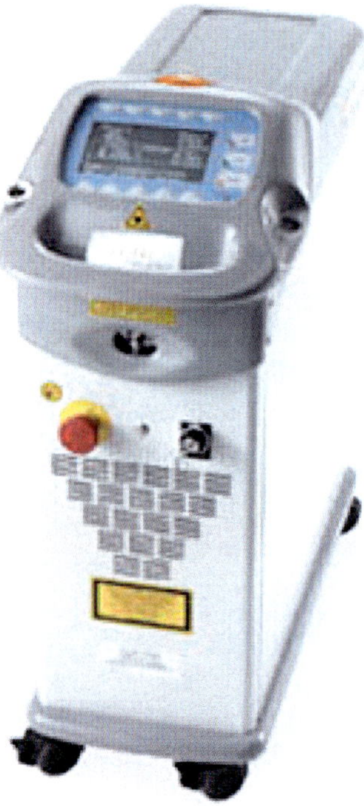

The Smartlipo Laser
Unit Workstation

The Smartlipo laser system is a new concept that combines all the latest generation laser technology with the advantages of compactness and reliability necessary for creating its operating performance. The micro-cannula with an optical fiber at the tip, conducts the laser beam in melting fat. Parameter adjustments are simple with intuitive, user-friendly software. By using a laser tip, the concept is relatively simple and is a superior and effective technique in lipo surgery.

The laser-tipped cannula emits power for extremely brief periods with electronic control of the emitted power given off by the laser pulses.

These light features can also emit thermal (heat) energy by concentrated and very brief pulses. The tissue treated with the laser beam is able to cool since the pulse wavelength only represents a tiny percentage of the time lapsing between two consecutive pulses. This highlights the technical capacity system that is able to deliver powerful, safe and instantaneous pulses.

For patients wanting the procedure, these highly technical insights may not be fully understood, but it is nice to know for safety and peace of mind, when you have Smartlipo performed by an experienced surgeon there are no harmful effects or burns produced by the laser.

In addition, the Smartlipo instruments have undergone multiple changes and upgrades. The latest laser, the MPX, is state-of-the-art technology. The latest model is energy saving and is managed much more efficiently by the operating parameters, while in their simplicity and anatomical design, the hand pieces and tips allow for carrying out effective treatments.

The surgeon, using the hand piece, can see the light beam glow at the tip of the cannula through the skin. This is much like holding a flashlight pressed against the palm of the hand in the dark and viewing bones on the opposite side of the hand. The laser light is conveyed through the micro-cannula's fiber that is 300 (with a diameter of 1 millimeter that is about the size of a half grain of rice) and may contain an optical fiber of 600 micrometers in size (about the size of a full a grain of rice). In short, the new laser instrument is efficacious and minimally invasive, ensuring rapid recovery time and reduced constrictive bandaging.

The laser-tipped cannula at 1 millimeter in diameter is smaller when compared to a traditional liposuction 4 millimeter cannula. It is less invasive, requiring only a small puncture of the skin to insert.

During the process of the laser rupturing fat cells, the cells may liquefy, leaving an oily liquid substance that is then removed for the most part by the surgeon through the same, tiny incision in the skin where the cannula was inserted.

As the laser comes into contact with the undesirable subcutaneous fat (fat that is stored in the body for future energy), the high peak power of the laser emission ruptures the membrane of the adipocytes (fat cells) and coagulates tissue as well as blood vessels in the fat.

When the laser process causes small blood vessels to coagulate immediately on contact, the result is less bleeding, swelling and bruising, and a more rapid recovery time than conventional liposuction. Because general anesthesia is usually unnecessary, the procedure is very safe with only minimal side effects.

By using the carefully calibrated 6, 10 to 18 Watt and MPX tips of the laser and with the infusion of tumescent anesthesia, liquefied fat deposits, when drained leaves an excellent sculpting work on the targeted area.

Here are the key points of the Smartlipo instrument and cannula in summary:

- The laser is introduced through a small cannula, delivering energy directly to subcutaneous fat cells—causing them to rupture and quickly drain away. The emitted energy also coagulates tissue, leading to tissue tightening.
- The laser causes small blood vessels to coagulate immediately—resulting in less bleeding, swelling, bruising and trauma (compared to conventional liposuction), for faster healing and recovery.
- Given the cannula's small size, Smartlipo is a minimally invasive procedure that usually requires local anesthesia—enhancing patient safety and reducing the likelihood of side effects.
- The laser provides a powerful visual guide during treatment.
- Unique pulse format permanently eliminates fat cells as well as coagulates tissue — resulting in tissue tightening.

Less Down Time

Many people who have this less invasive laser treatment are back to work in three days or less. There may be some bruising and swelling, but these areas are often covered with clothing.

One patient had this comment:

"After I had my tummy flattened, in roughly 24 hours after surgery I was up cooking breakfast, changed my sheets and washed a load of laundry. I am sore and walking slightly bent over but I feel good."

The Response to Smartlipo

In a recent poll, 90% of respondents that had the Smartlipo surgery said that it was worth doing. Even so, Smartlipo is not a solution to obesity. It is designed to eliminate fat in specific areas of the body. Ideal candidates are those that are in relatively good health mentally and physically and who are not excessively overweight.

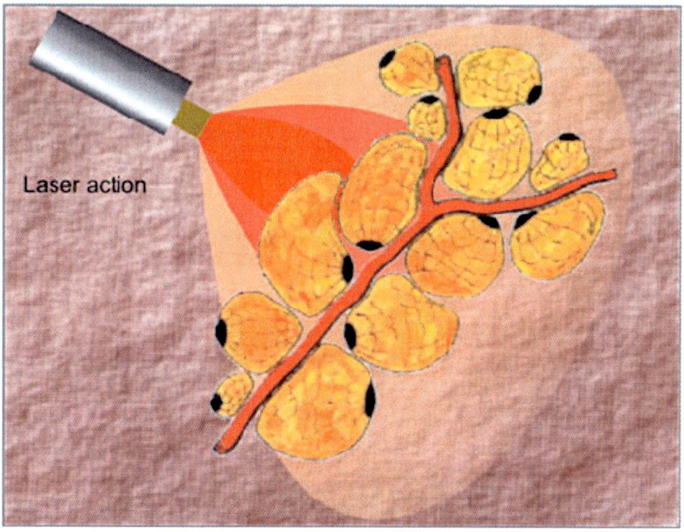

The small cannula with the laser tip in action, melting the fat cells in the dermis just beneath the skin is very effective. Notice that the final result after applying the laser leaves little fat remaining.

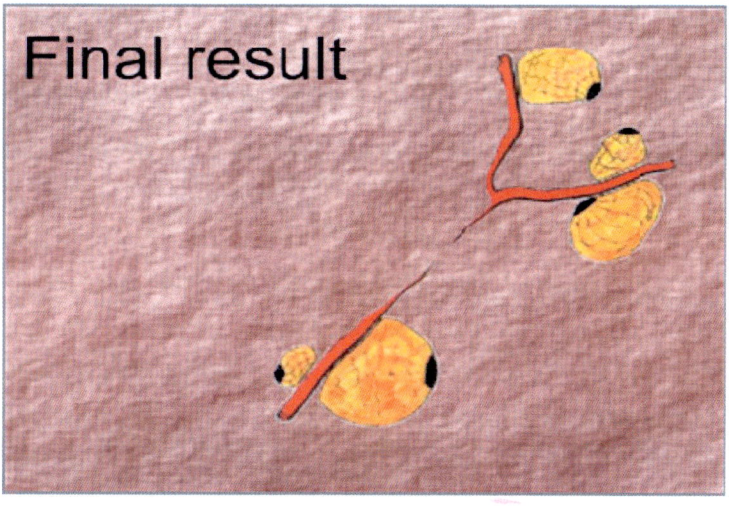

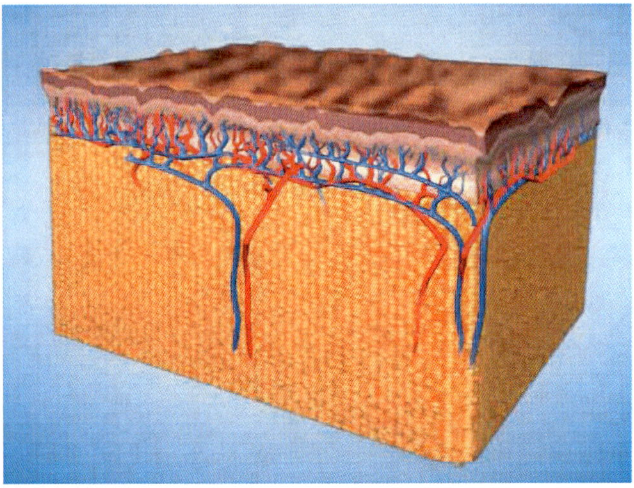

This is a cutaway of human skin and fat cells beneath the outer layer of skin, with blood and nerve endings. Vessels and nerves are severed with the laser, but quickly reform and connect in an amazing healing process of the body.

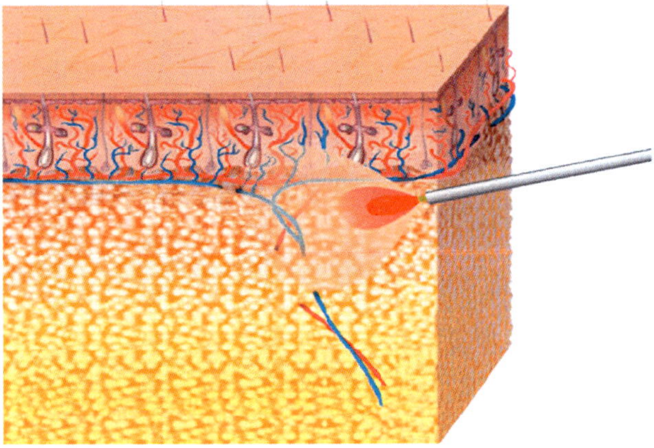

2

"If you can pinch a substantial amount of fat,
I can remove it."

Fat in Areas of the Body

You've adhered to eating right and exercising, yet you still can't lose those "love handles" or other fat deposits you consider unsightly. We all know that fat, or adipose tissue, can be very difficult to get rid of through exercise and diet alone. It can be resistant.

In those delicate areas of fat deposits, such as small areas of the face and jowls, with an experienced surgeon fat removal results are possible. There may be several other areas of the body that need sculpting as well: the neck, underarms, abdomen, love handles, inner thighs, calves, knees, and denser areas such as the male breast. Smartlipo laser fat removal is a very effective sculpting method.

Many researchers of adipose tissue (fat) insist that adults have a fixed or constant number of fat cells in the body. Because liposuction permanently reduces the number of fat cells in treated areas, the improvement is permanent, provided that stable weight is maintained since the fixed fat cells have been removed.

We all know from experience that fat in most adults tends to increase gradually over the years. After the age of 30, many individuals tend to add fat according to a genetically predetermined pattern. An example of a genetically predetermined fat distribution that is resistant to dieting and exercise is fat that bulges on the abdomen and hips of a woman after pregnancy. The abdomen may also be an issue with many men.

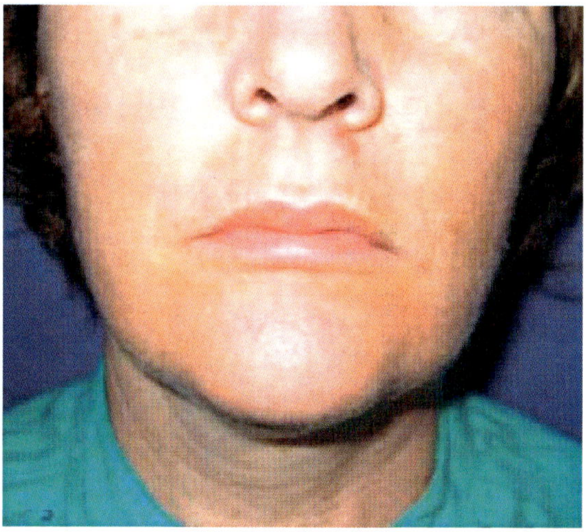

The jowls before Smartlipo.

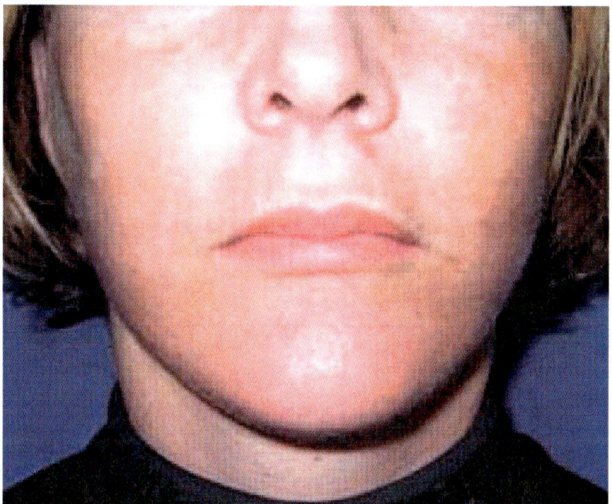

The small amount of fat in jowls gone
after Smartlipo.

Most people are not pleased with certain areas of their body that bulge with fat. Even those who carefully watch their diet and do physical exercise are often troubled by unwanted areas of fat in regions of their bodies that they simply can't seem to shed; this especially so with the tummy, hips and waist even after months in the gym. These areas are commonly cleared of fat with the Smartlipo sculpting process which often creates a lean look, beginning with the cheeks and descending all the way to the ankle combined with exercise and diet shows remarkable improvement. Smartlipo is effective in small areas — cheeks, double chins, turkey necks, jowls, arms, knees, saddlebags, and enlarged breast tissue in men and other areas of the male or female body. It is the solution for those who are not excessively obese.

Cheeks, Jowls and Neck

Cheeks with excess subcutaneous fat can cause a person to appear chubby and is often the result of genetics in the family. If you are bothered by excess fat in your cheeks, there may be a solution with Smartlipo that can reduce the excess fat. The amount of fat in most people's cheeks is very small and therefore traditional liposuction is not normally applied. Once again, diet and exercise can help many areas of the body, but for some, nothing seems to affect those fatty deposits in the cheeks.

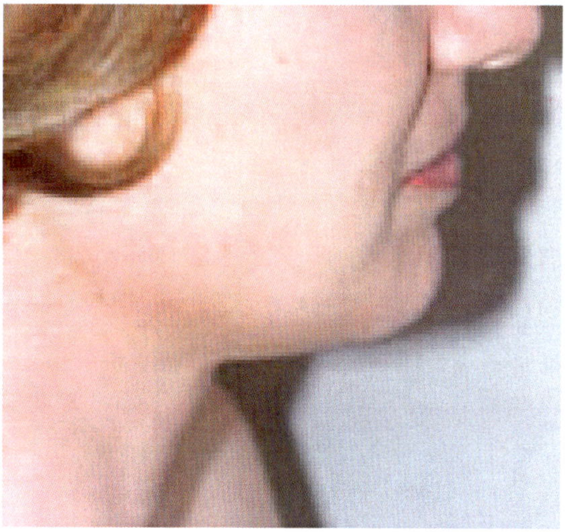

Before chin Smartlipo

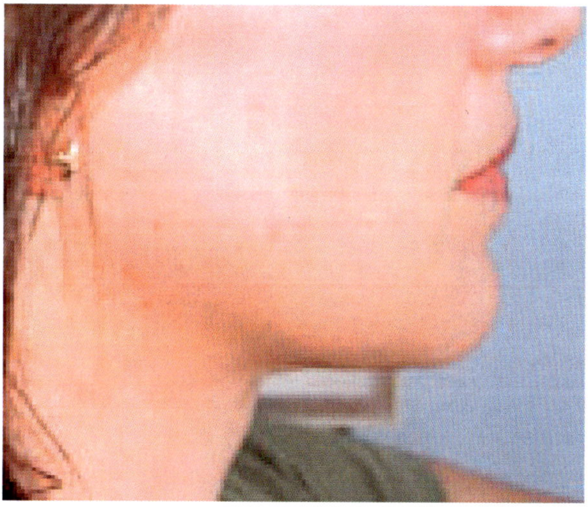

After chin Smartlipo

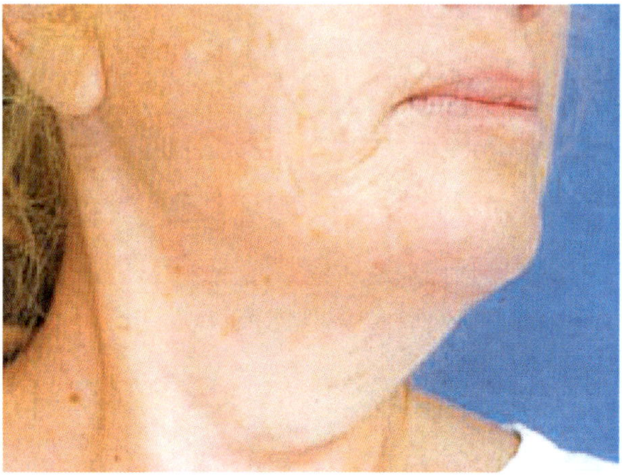

Before chin/neck Smartlipo

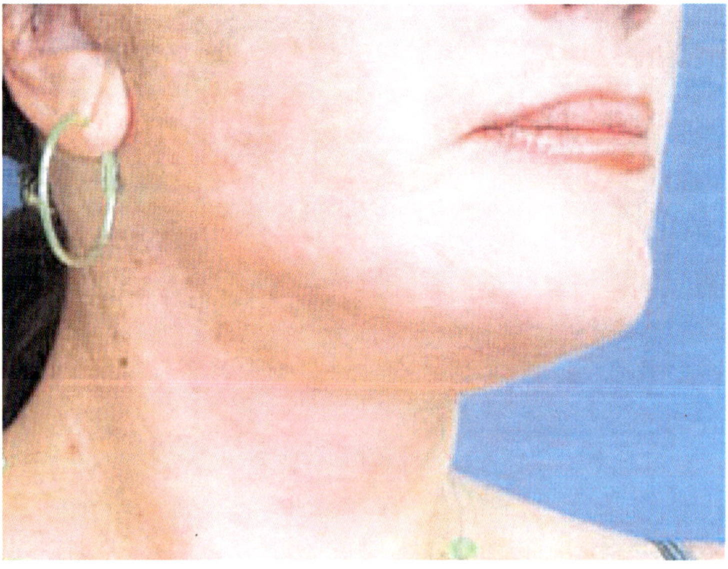

After jowls/neck Smartlipo and the laser tightening

During the surgical procedure, the small cannula is most often inserted through a tiny incision near the base of the ears to conceal any possible obvious scarring. It is worked along the cheek and Jowls to melt the fat and leave a thinner layer of skin along the hollow areas of the cheek allowing the jowls to lie closer to the jaw bone in a very pleasing, sculpted appearance. It will heal with little bruising and soreness in a few days and with many patients, the skin will tighten and look more youthful.

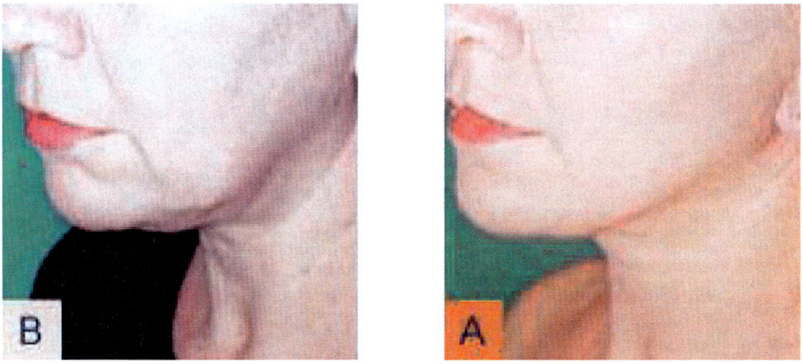

Before and after Smartlipo of the jowls and slightly drooping chin

Smartlipo significantly improves the contour of the neck, jaw line and cheek area. The procedure is useful for patients with good skin tone who have excess fat beneath the skin. Results are dramatic, with tiny hidden incisions, and recovery is rapid.

There is usually a minimal amount of discomfort following the procedure. Most swelling and bruising will resolve in a week. Keeping the head elevated and applying cold compresses helps to keep the swelling down.

Upper Arms

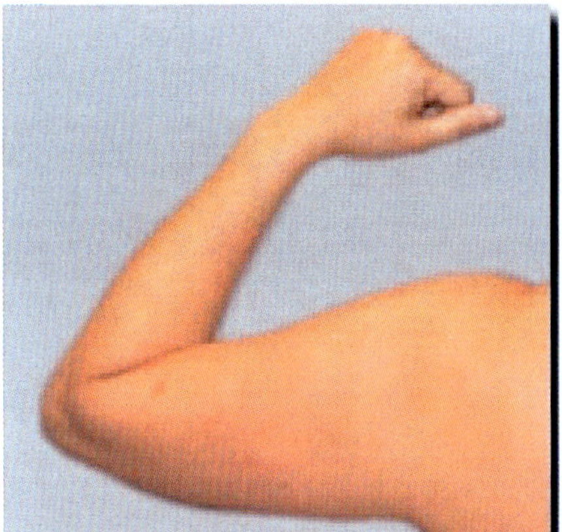

Before

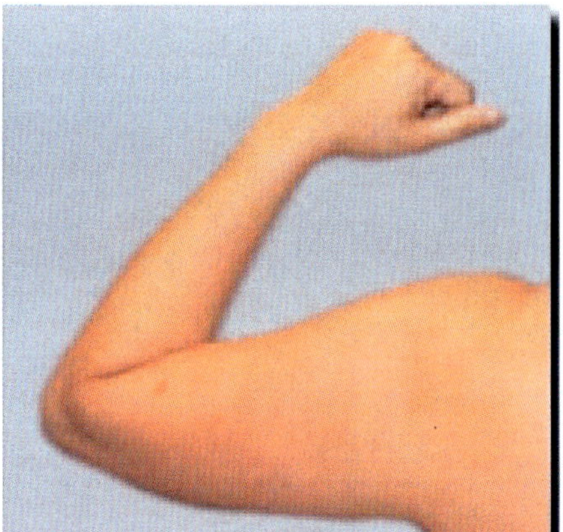

After

Male Breasts

While sculpting the fat with the smartlipo laser, the fat beneath the areola (brown skin around the nipple) is melted with attention to the amount of fat remaining to give a full, pleasing appearance. The amount of breast tissue remaining after treating gynecomastia with Smartlipo is minimal, though it is one of the most difficult procedures of fat removal due to the combination of fat and tissue.

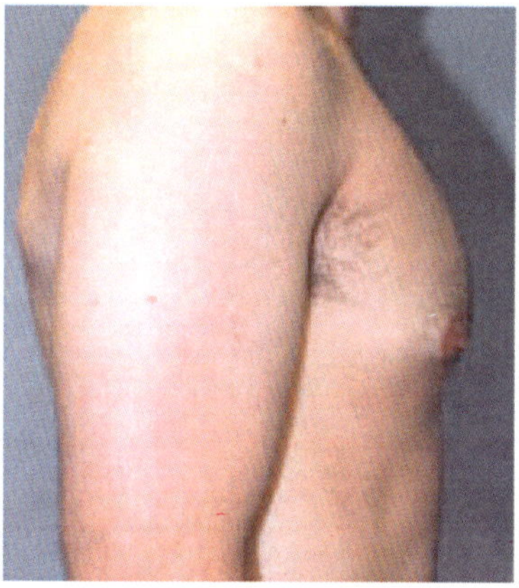

Before

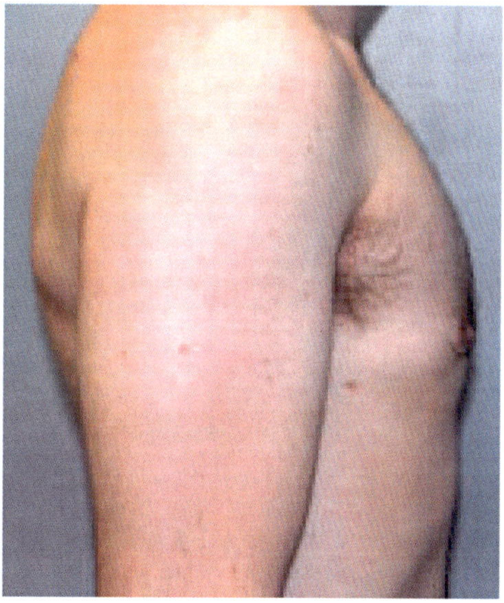

After

Abdomen

The abdomen is the body area most frequently treated by Smartlipo on a patient who has a protruding stomach and wants it flatter. A number of men and women have excess fat in either an upper or lower abdominal area, and some have both. Abdominal fat deposits are often extremely resistant. It is for this reason that people who diligently try to tone their bodies become frustrated with a protruding abdomen. Fortunately, moderate amounts of abdominal fat respond well to Smartlipo. Those fat cells are eliminated and will never return. Of course surrounding cells can increase in size (through overeating) and fill the space.

While taking out the fat in the abdomen, the surgeon must do upper and lower if there is to be a satisfactory result. The surgery, with some, can result in a "six pack."

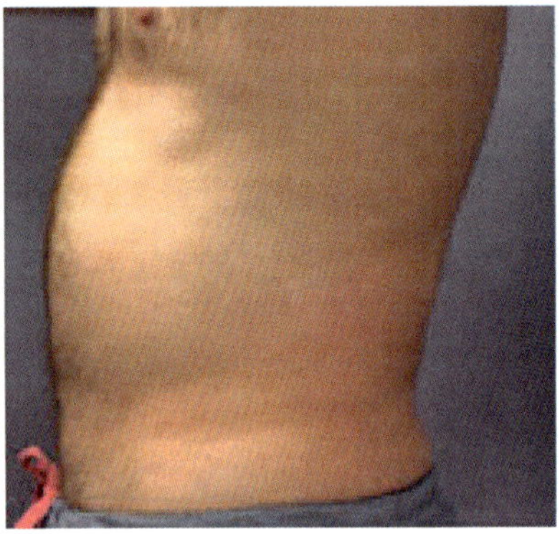

Before

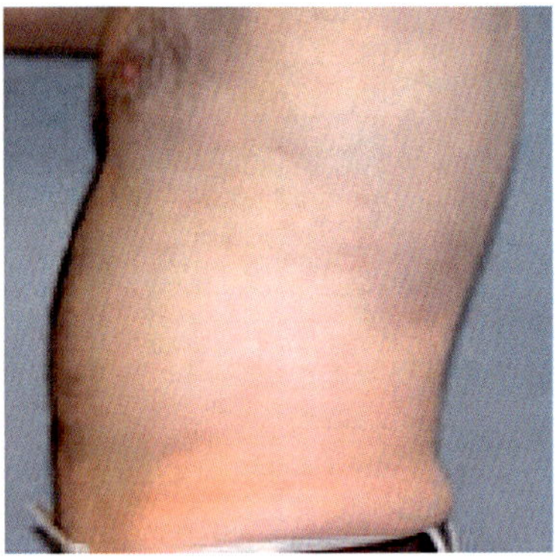

After

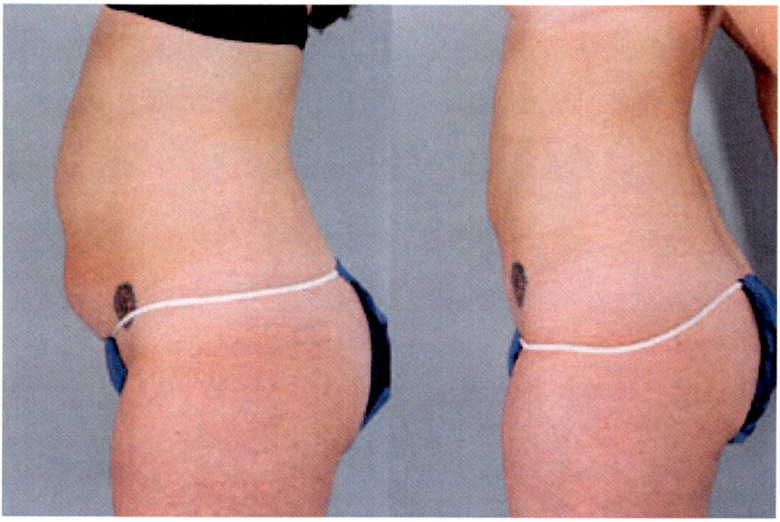

Smartlipo of the abdomen

Flanks (from back to side of spine)

Fatty deposits on the sides of the body just above the flanks are usually referred to as "love handles" or a "spare tire." No matter what name you give it, it is still a layer of fat that is put on around the midsection, especially visible on the sides. Love handles are more common in men than women, because the torso is one of the first areas of fat build-up in the male body.

Love handles are resistant to dieting and exercise. People who carry excess weight around their midsections know that it can be extremely challenging to get rid of that "spare tire," even with the most diligent regimen of sit-ups and dieting. For this reason, Smartlipo is a popular option for eliminating love handles.

This fat removal by laser can be effective with some women who have gained fat in the flanks due to child birth or have a genetic issue. It is possible to regain those feminine curves in the waist with the laser.

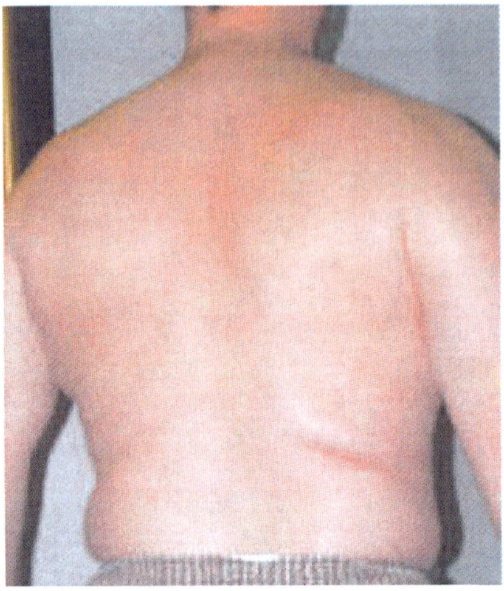

Patient in late 30's received fat removal in the upper abdomen, and waist.

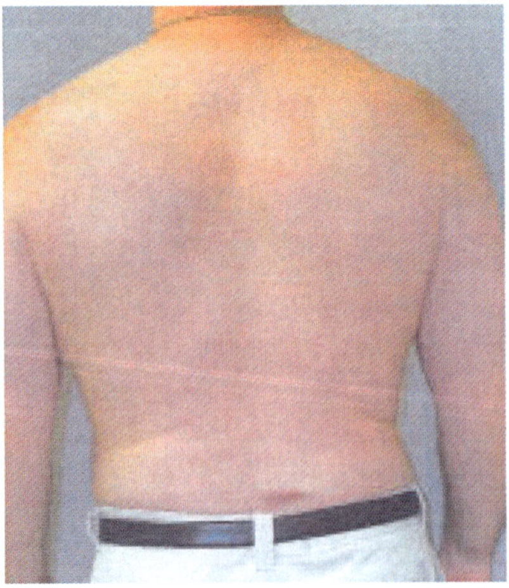

After removal of fat

Hips and Saddlebags

Smartlipo is especially effective with slightly protruding hips. Females tend to have more fat accumulation on the hips than men. Women retain fat along the hips even when they are clearly not overweight. Curvy hips are a hallmark of pleasing appearance on women, but so-called saddlebags are not welcome.

Smartlipo is generally successful in creating a more contoured figure on women who have a slight amount of protruding fat with the saddlebags. This procedure on women with mild pockets of fat along the hips may be performed with a local anesthesia.

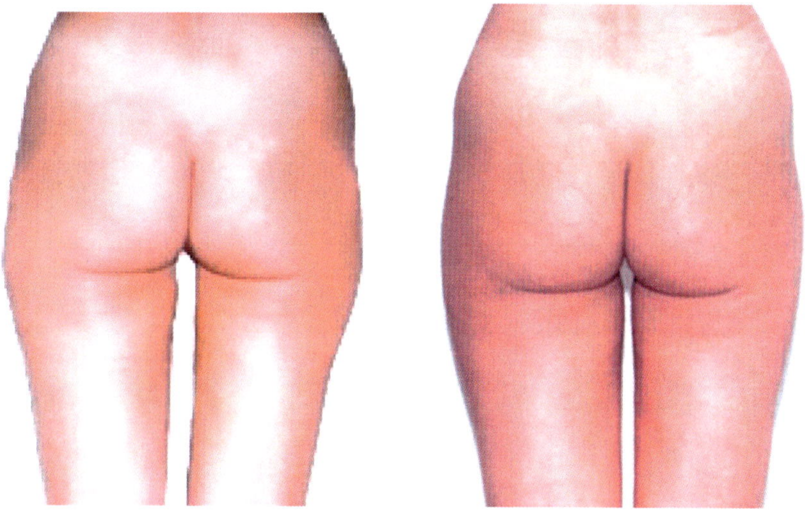

Smartlipo removed the slightly bulging saddlebags.

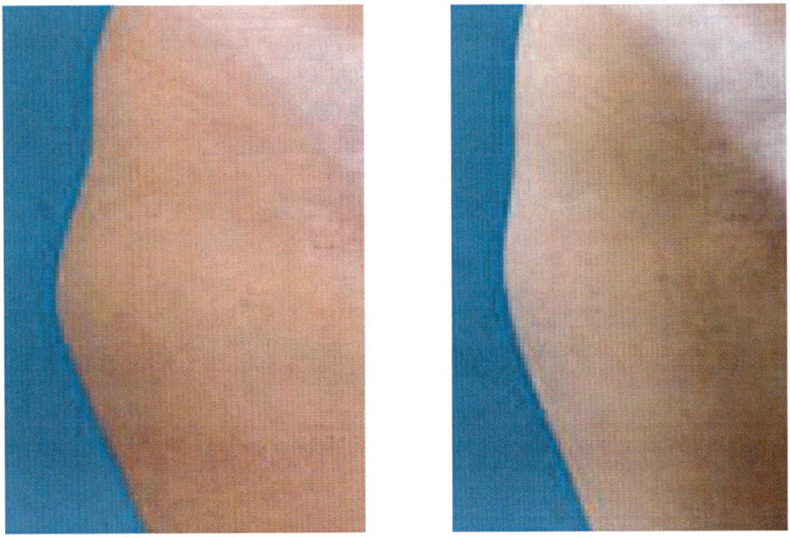

Saddlebags from the front removed with Smartlipo

Cellulite Removal

Though Smartlipo is not specifically designed for removal of cellulite, it does have a positive success record with removal of the cottage cheese appearance. As the cannula probe melts away the fat, the skin tightens. In most cases there is a noticeable diminishing of the dimpled appearance in the skin.

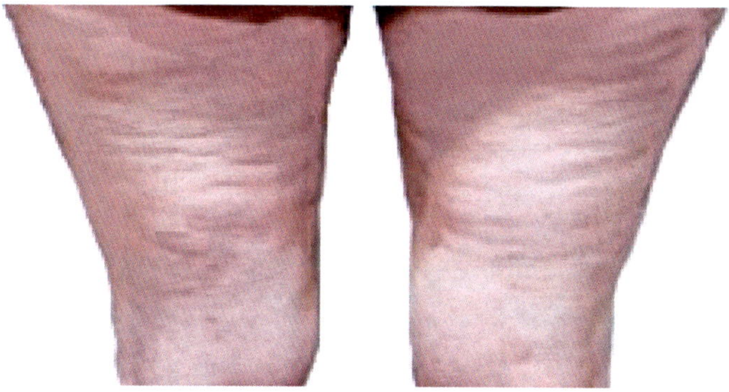

Cellulite Removal

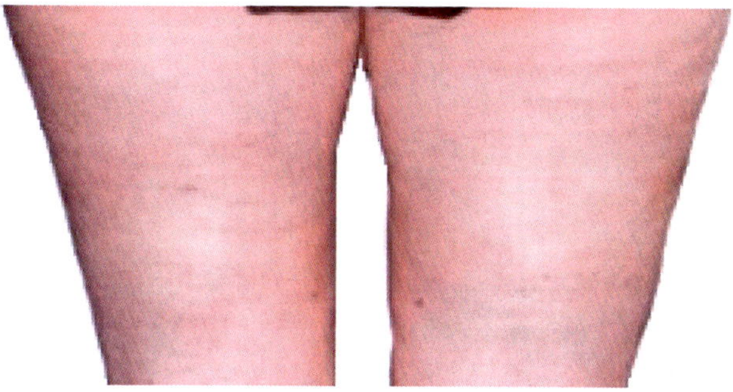

Conturing Around the Buttocks

While men gain weight around the waist, women tend to have an increase of fat around the buttocks. Reshaping this area with the laser can offer a gratifying result with many women.

The issue is usually the sacral fat pad that bulges with fatty tissue or the flabby creases at the lower area of the buttocks often referred to as the "banana" shape. This is one of the most difficult areas to obtain sculpted results.

These areas can become unsightly with excess fat. There is also the horizontal plane of the buttocks that many want more defined.

But this does not matter. You may be dissatisfied for several reasons and want more shapely buttocks. It is possible. Whatever your reason or goal, liposuction can provide stunning results that will leave you smiling. Make sure to have the buttocks reshaping procedure done by a surgeon who has an artistic touch.

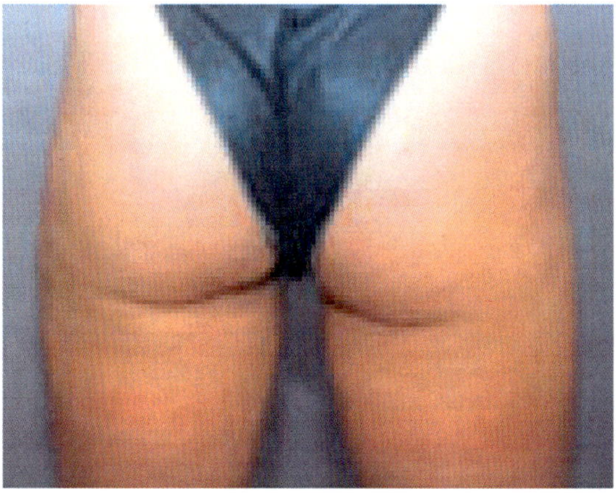

Before Smartlipo showing the slight bulge in outer thigh and under the crease where the banana shape is located

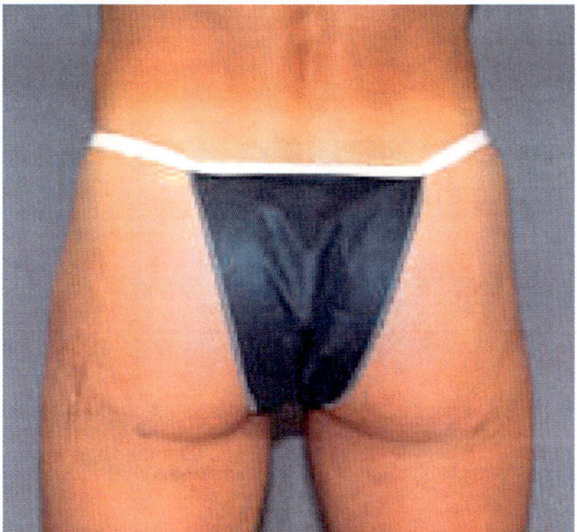

After fat removal

Thighs

Removal of fat in the legs, from knees to groin, and outer thighs and outer legs, from knees to waistline, may eliminate cellulite from this area of the body. The inner thigh is a particularly difficult area that requires skill to properly remove the fat to gain a slenderizing appearance. One of the most common sites for Smartlipo is the thigh. Here again, Smartlipo performed on the thigh can reduce areas of fat that have not responded to diet and exercise. Removal of fat on the thighs gives a silhouette of beauty to the lower body.

Knees, Calves and Ankles

There are those who are unhappy with the overall appearance of their knees, legs and ankles where fat settles causing pudgy knees and thick ankles. Since the knees, calves and ankles are highly visible, women want them to be shapely. Legs are characterized by a gradual tapering from the thighs to delicate ankles. Muscles and bones shape the legs and feet, making this a relatively difficult area to sculpt. However, it is possible to sculpt them into a shapelier appearance, though not all surgeons perform this difficult-to-sculpt area of the body.

Candidates for Smartlipo

Almost anyone can be a good candidate for Smartlipo. In general, the ideal candidate ought to meet the following requirement:

- You are in good health, routinely exercise and are within 25 lbs. of your ideal weight.
- You have one or more areas of local body fat deposits.
- You are bothered by the fatty areas of your body and are desirous of eliminating the fat permanently.

Ideal candidates are in good health, of normal body weight, and have realistic expectations for the results.

It is basically your decision. Most surgeons will perform the procedure if you meet the three requirements listed above.

There is no definite age or weight limit. Some of the happiest patients are within twenty to thirty pounds of their ideal weight, and have been stable at that weight for years.

Costs Associated With Smartlipo

The cost for having Smartlipo will be discussed during consultation with the doctor. This varies in regions of the United States depending on whether the doctor's practice is in an urban, suburban or rural area. The cost of surgery may include the physician's fee plus other costs such as the operating room, medications, lab tests and assistants. Be sure to ask your surgeon if your total costs cover all the usual, expected fees and charges and is a comprehensive figure of the complete package of having Smartlipo.

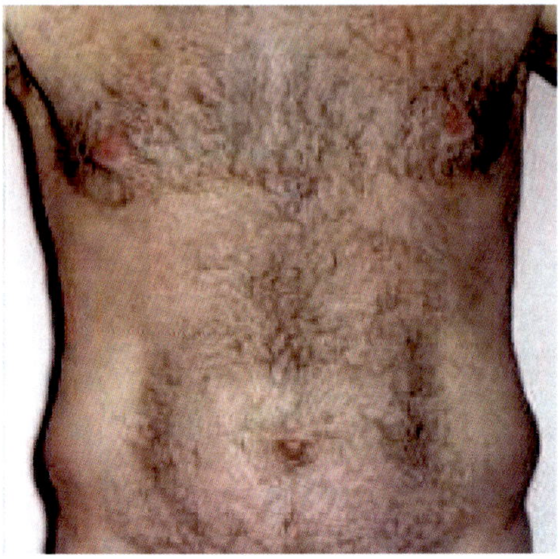

Minor sculpting of the slightly protruding stomach love handles. This patients looks as if he had spent time at the gym.

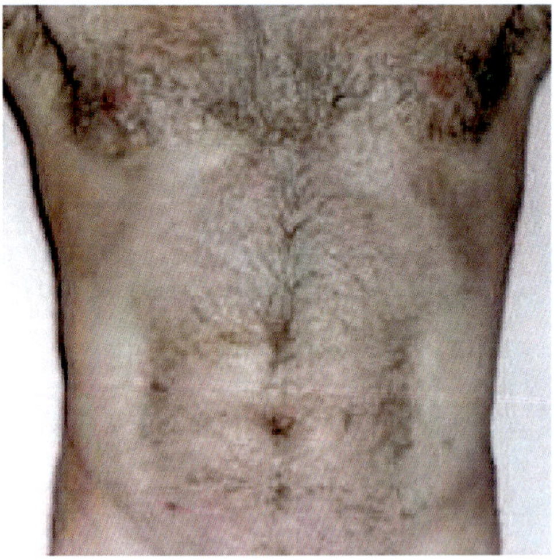

3

We've Seen Smartlipo's Amazing Results. It can change your life for the better

Overview of Results with Smartlipo

The advantage certain patients have with Smartlipo is the procedure's marvelous method of removing fat after many attempts of ridding the body of this non-responsive fat accumulation. Smartlipo is often a subtle removal of fat for those who may otherwise display a fit and toned body. The candidate for lipo sculpting could be troubled by a slight double chin, mild protruding abdomen, love handles or saddle bags. The person dealing with a slight flaw associated with fat is sometimes driven to have it align with the rest of their body. This is to be expected. There are those whose bodies appear acceptable to most of us, yet they still want that "babe body" and seek after it to near perfection.

Of course this is not always the case. Smartlipo, as you will see in the photos, can remove large regions of fat, such as the abdomen, the hips, and the thighs. More and more patients are opting for this less invasive method of fat removal due to less pain and a faster, shorter recovery.

Here is a review of successful Smartlipo fat removal with several patients.

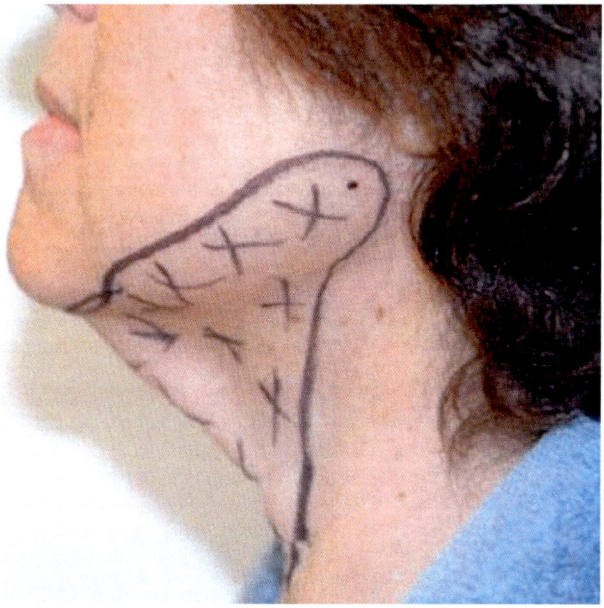

This before photo shows the markings where the laser will sculpt the neck and jowls.

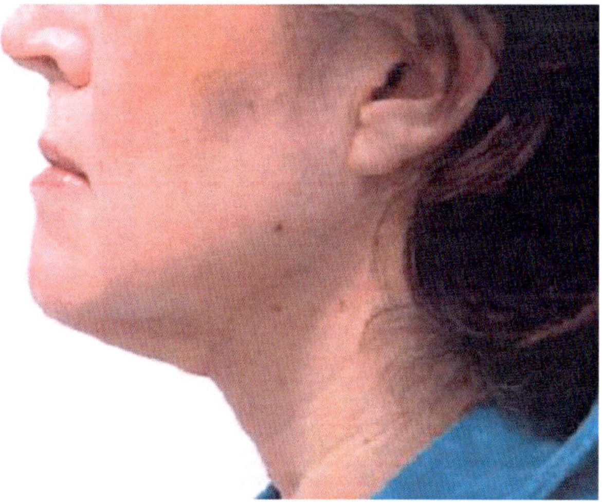

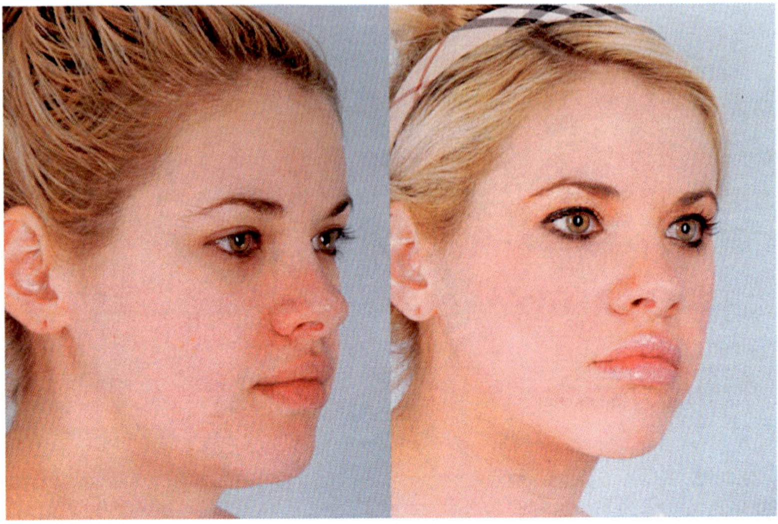

This teenage girl, after laser fat removal, suddenly looked like the Prom queen.

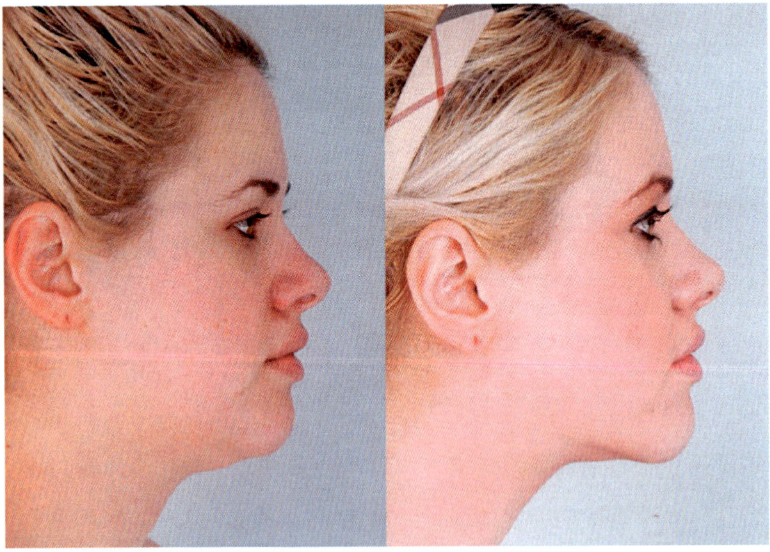

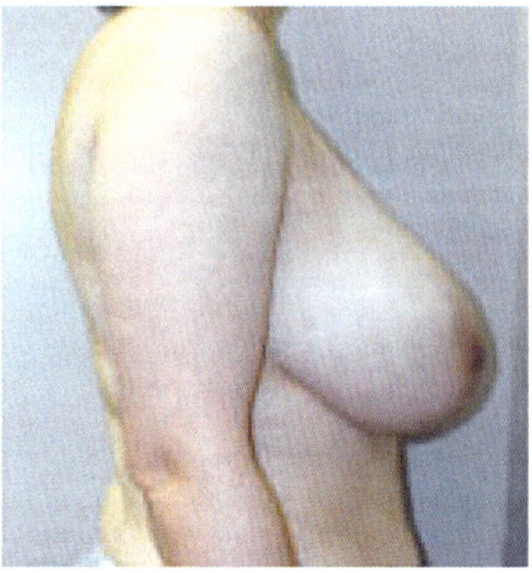

Before

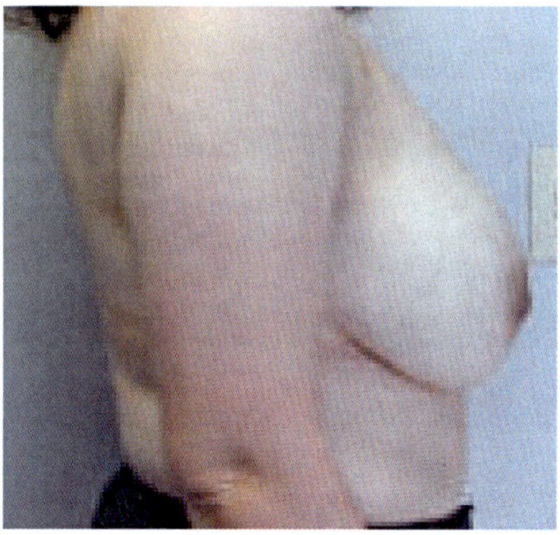

After

Patient in her 40's received a breast reduction using Micro-Lipo techniques. This brought her breast size down from an F to a D without the need for removing excess skin through a breast lift.

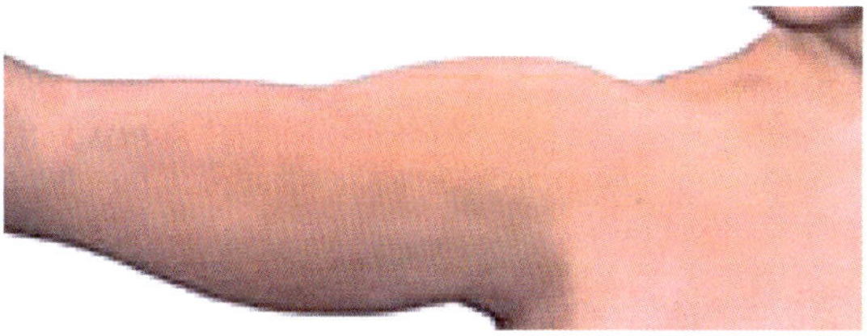

Slenderizing the under arms.

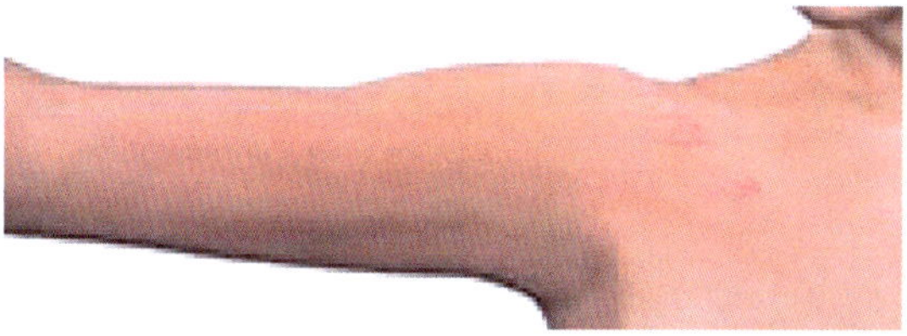

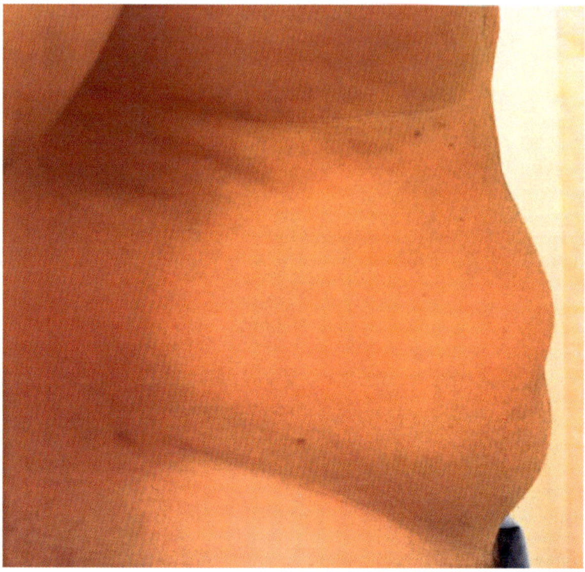

Sculpting the upper and lower areas of
this patient's abdomen

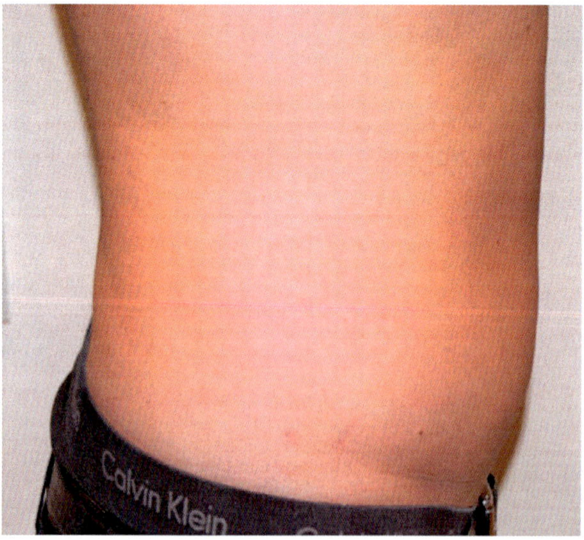

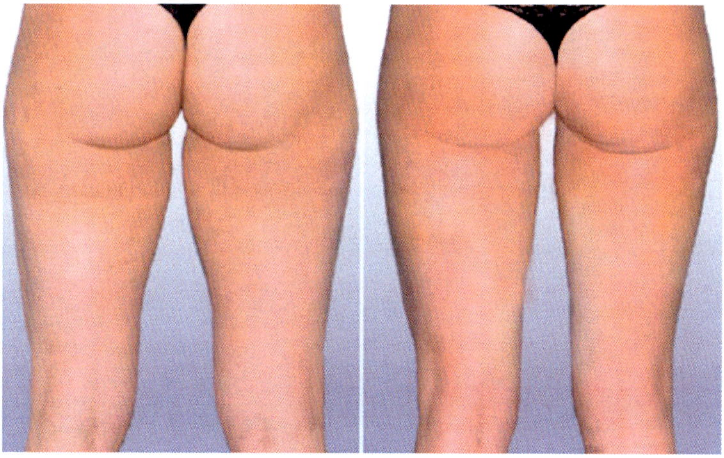

Results of two patients each wanting a more attractive body from the hips down

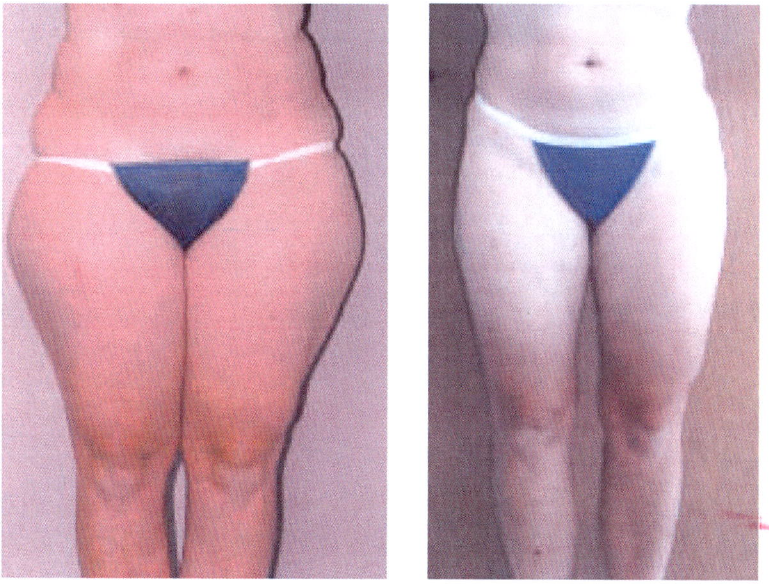

This young man wanted his chin and neck sculpted.

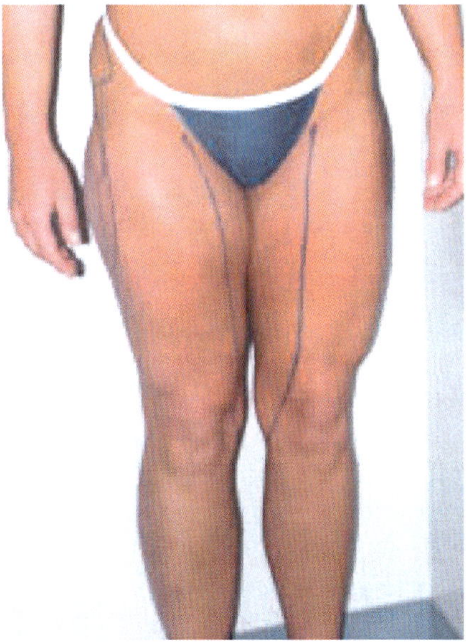

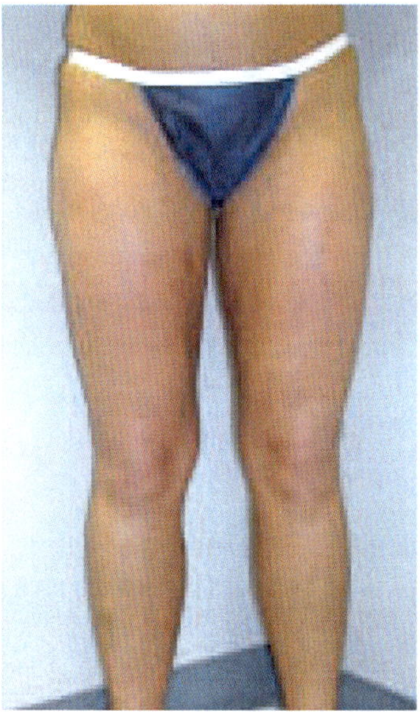

Here is a 22-year-old female who is very fit but wanted sculpting with Smartlipo. The Smartlipo procedure took place on the upper and lower abdomen, lateral waist, flanks and hips.

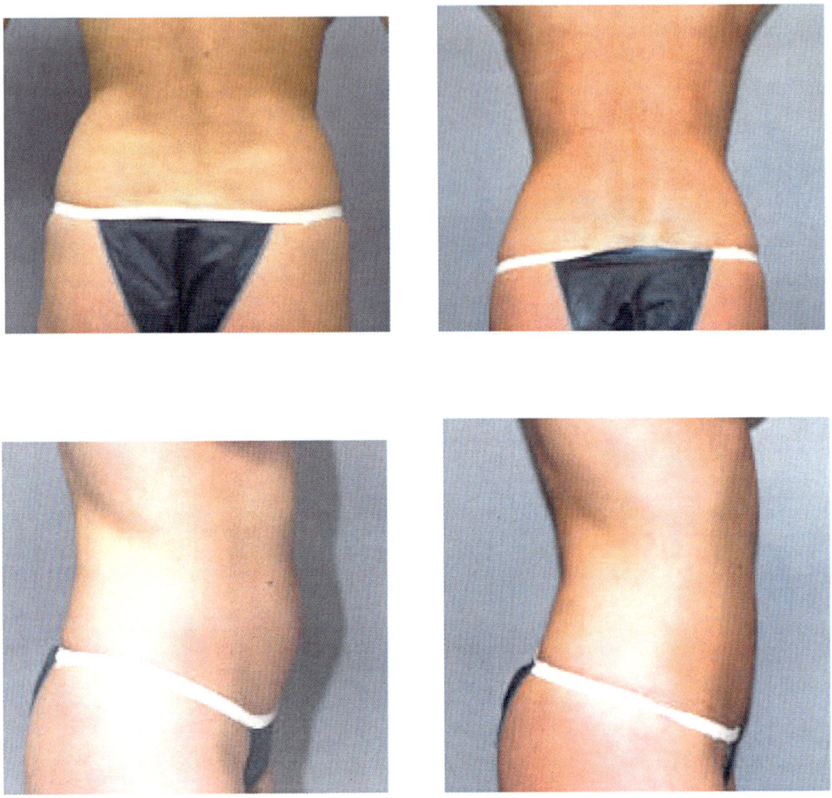

This 33-year-old female wanted Smartlipo and improved abdominal appearance. The procedure was discussed in detail as Dr. Schafer performed Smartlipo of the upper abdomen, lower abdomen, lateral waist and hips.

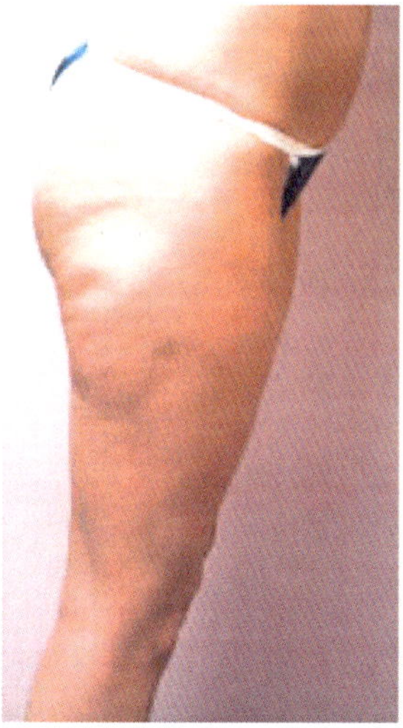

Before

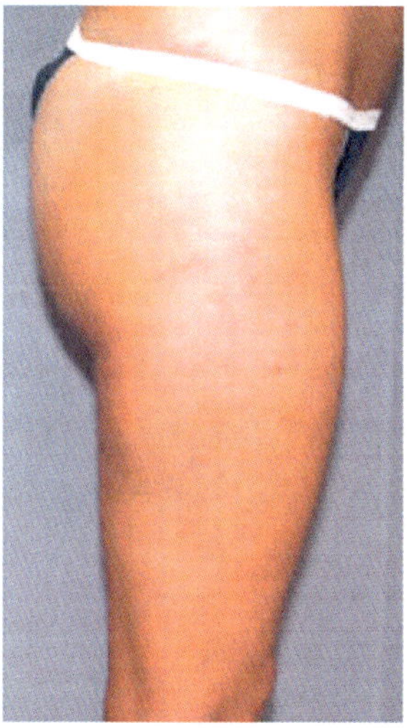

After

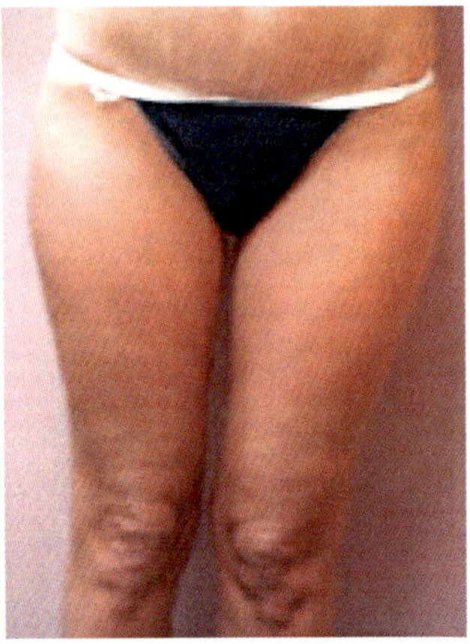

Before

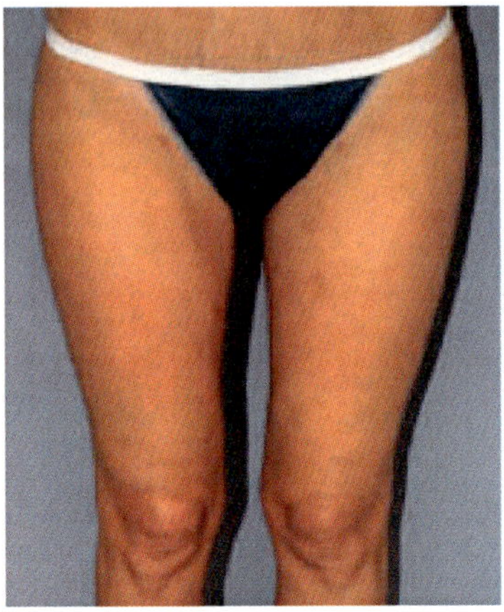

After

This 42-year-old woman is in excellent shape but unfortunately was cursed by the genetic trait of cellulite in the legs. She tried Endermologie and other treatments in the past and did not see any visible results. Although there is no guarantee that the cellulite will improve after the procedure, historically a significant change is visible. The patient had Smartlipo of the inner, outer and back (posterior) thighs. The results are remarkable and will vary patient to patient. After photo, shot less than one month following surgery.

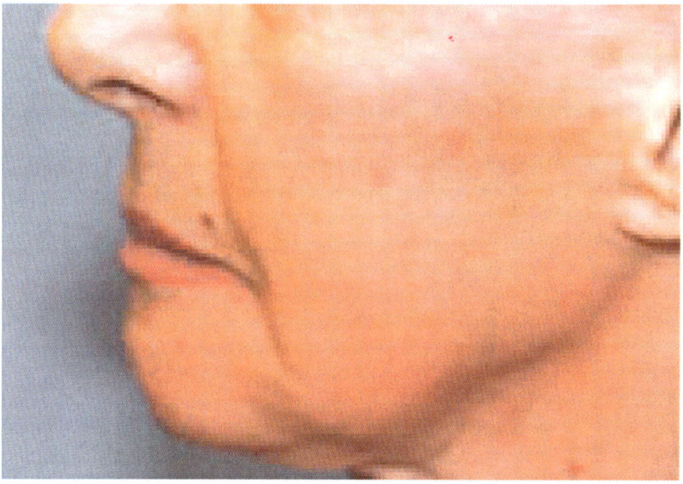

This is 61-year-old woman from San Diego. Being native to the area and after years of sun damage, she decided to have Smartlipo in the neck, along with a laser lift and laser resurfacing.

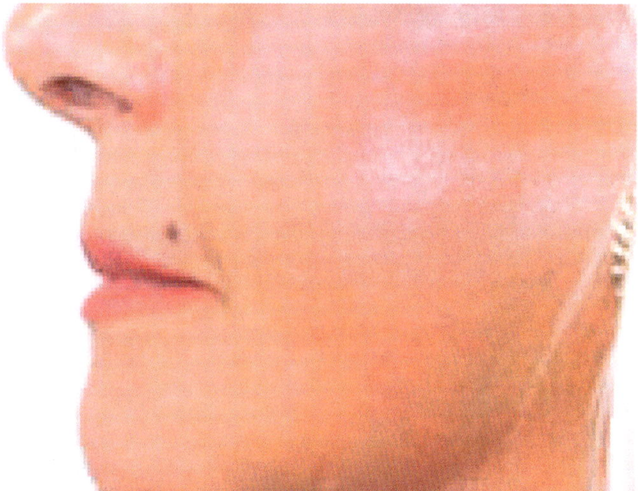

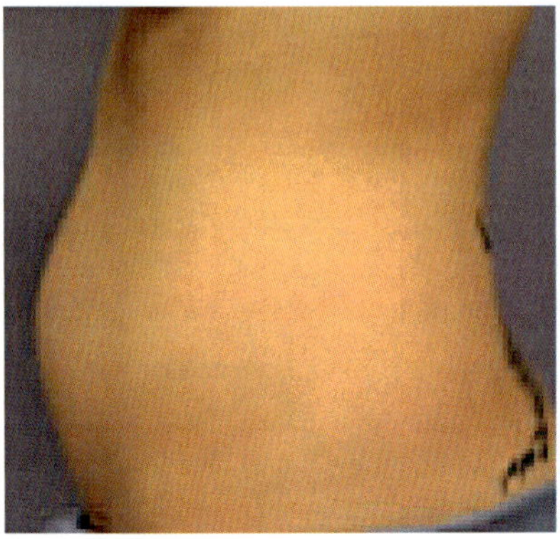

This is a 30-year-old patient who is clearly in excellent shape but has had problems getting rid of her baby belly. She had Smartlipo performed on the lower abdomen, lateral waist and flanks. She looks great.

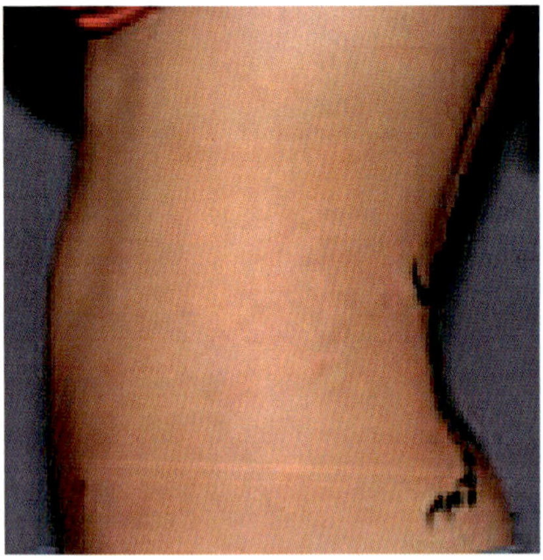

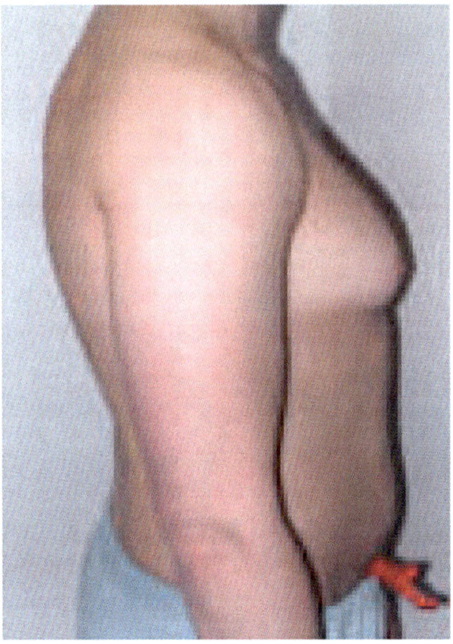

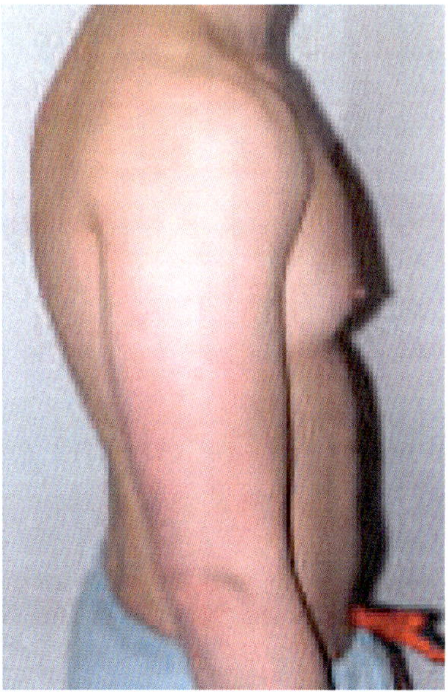

This male patient in his 30's received Male Breast Reduction surgery and microlipo of the upper and lower abdomen. He didn't want anyone to know about his procedure and since the scars were so small, no one has noticed. The patient is much more comfortable with his shirt off and feels more masculine.

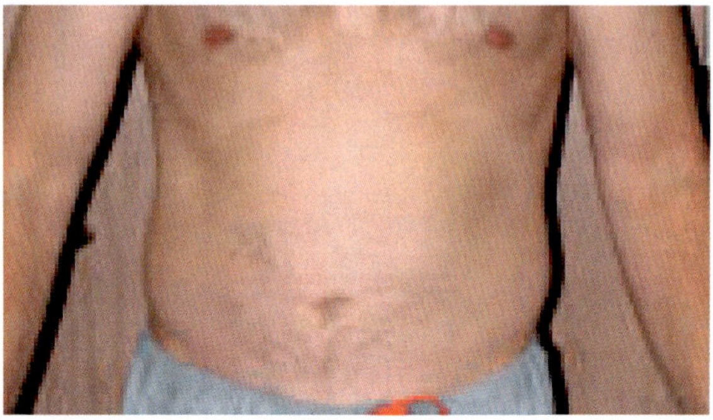

Patient in his 30's received Smartlipo of the upper abdomen, lower abdomen and waist.

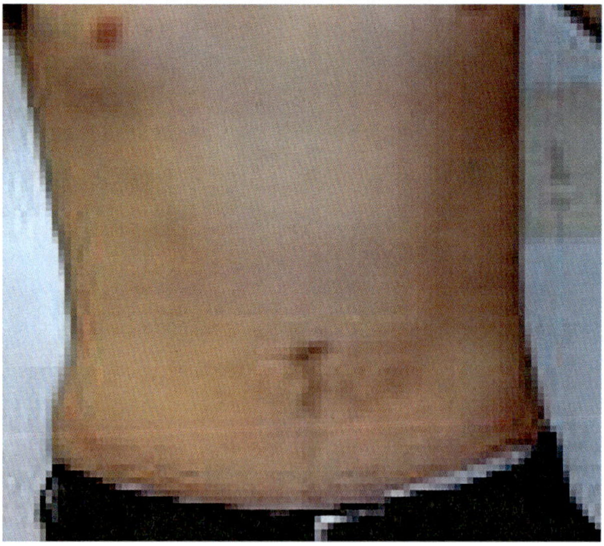

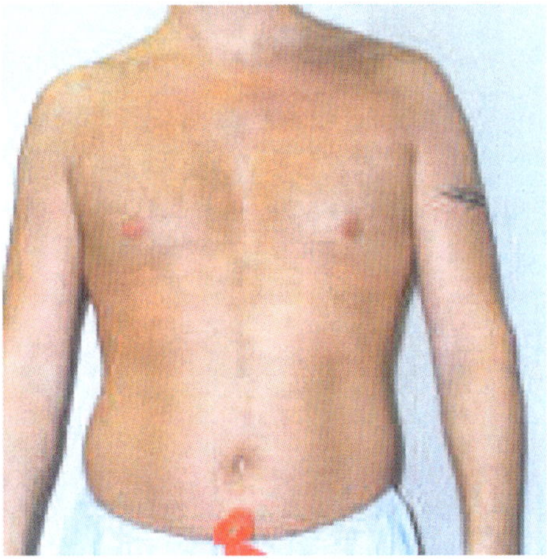

Although this is what many men want, the time and effort that go into toning the body to create this sculpted look is an enormous commitment. The laser did the trick.

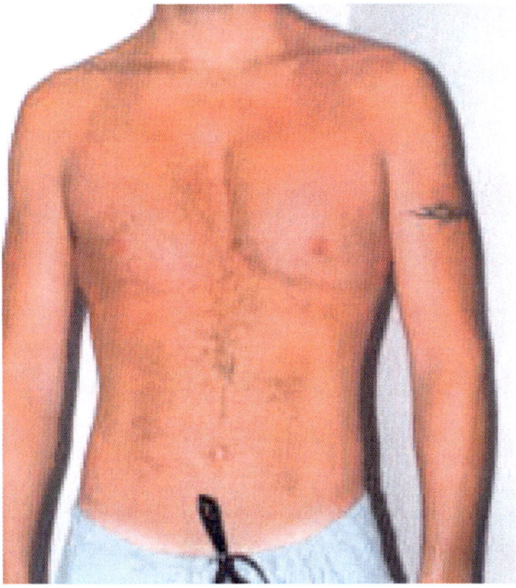

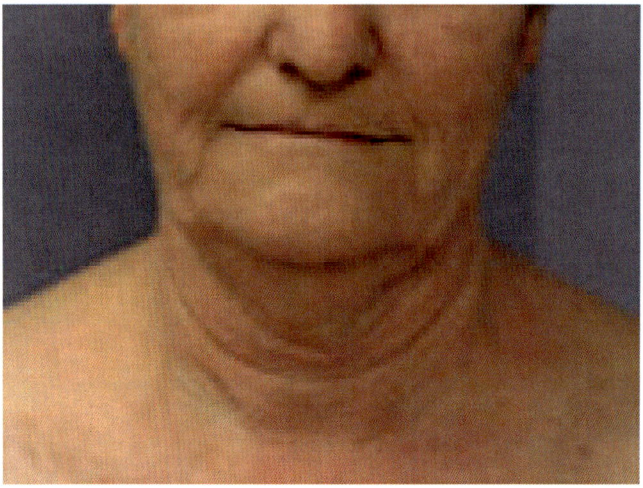

Here is 61-year-old woman from San Diego. Being native to the area and after years of sun damage, she decided to have Smartlipo in the neck.

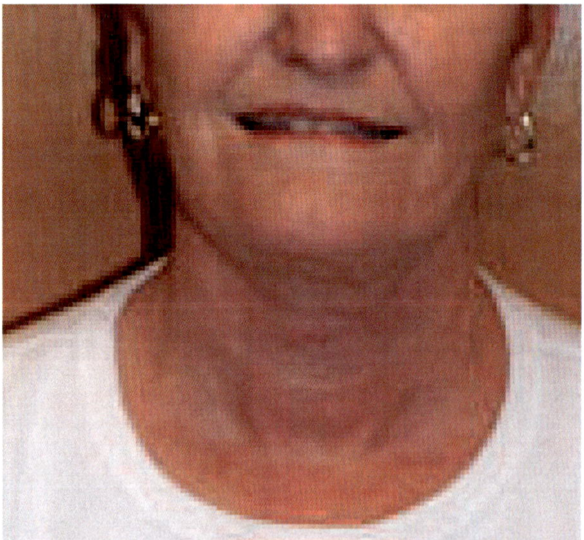

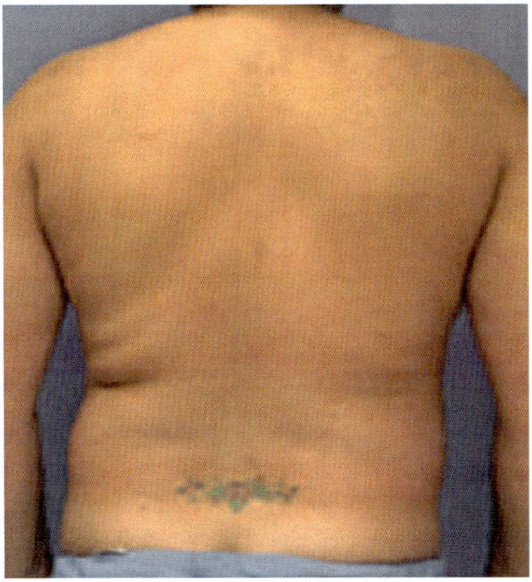

Before

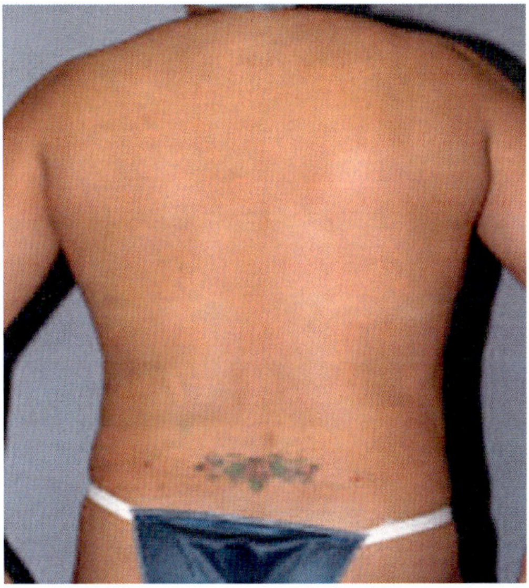

After

Pre- and Post-Op Personal Checklist

- First and foremost, follow your doctor's instructions. The doctor who has done a number of Smartlipo procedures knows what is best for your recovery. Listen to the doctor.
- Take a few days off from work. Your doctor can best advise you on this during consultation.
- Have someone drive you home or to your hotel from surgery.
- Have meals prepared ahead of time, if this is one of your duties.
- Fill your prescriptions as directed
- Prepare your bedroom with items close to the bed that you may need to take, such as medications, grooming things, tissues, water, handwipes, reading material, remote clicker for tv. Straws or sippy cups are handy little items to have nearby. Think ahead as to what you may need while resting in bed for the day or two you are recuperating. Have a bell nearby to signal others. Some even have a baby monitor in the room. If you have one, use it.
- Keep your doctor's phone number nearby and also the numbers for emergency.
- Use anti-bacterial soap.
- You need a night light in the bedroom and maybe in the bathroom.
- Drink Gatorade the first day after surgery. Have a low-salt diet for two to three weeks thereafter.

4

Smartlipo is the least invasive fat extraction surgery in the world today.

Consultation Laser Surgery and Recovery

Consultation

Anyone considering the new laser fat removal procedure wants great results. If you feel comfortable and confident that you have selected the right surgeon to do your Smartlipo procedure, the next step after making an appointment is the consultation with the doctor.

There several point to consider when you make your appointment to meet in consultation Know what you want and think it through. Be realistic. There is a limit as to the amount of sculpting that can be done. In most cases there is a dramatic difference when the body fully recovers from the melting of the fat and it is not unusual for people to go down at least ten sizes.

One of the most challenging aspects of Smartlipo is making the decision to proceed. You have to ask yourself if it is really what you desire. Obviously it ranks high on your list for improving your body's appearance, but you must first decide if you want to get rid of problem areas that are prominent in your body contour.

The best way to make a decision is to gather as much information as possible and then proceed to meet with the doctor in consultation. If you have concerns talk about them. A reputable and honest surgeon will give you frank answers to help in the decision.

Listen to the doctor and make a decision based on what is gleaned from the consultations. You are the one who has to live with your decision. Take counsel from those you truly respect. Stand firm. It is your body. It is not a good idea to have surgery because you are depressed. You need to have an exceptional level of positive thinking. The best way to gain this is to look forward to becoming both physically and mentally strong.

For most people Smartlipo can be a quick fix type of cosmetic surgery, but the long lasting results will only endure as long as you are willing to control the food you put into your body and exercise. Each decade your metabolism slows down. You either need to exercise more or eat less. Smartlipo does not fit the classic description of the premise of "going under the knife." This can be a scary phrase that is not part of the Smartlipo procedure.

The tiny incisions that are made allow the cannula to enter the fat. They are so minor that in most cases there is no need for sutures. The procedure is very mildly invasive. However, caution must be considered. The phrase "It is just liposuction" should not be taken lightly.

If you don't want people to know you've had fat removal, they will not. You can arrange with the surgeon to have a private office visit and recovery. Perhaps your surgeon will have other suggestions to preserve your privacy to the level you desire.

If someone asks you if you have had some sort of fat removal, it is your decision. One consideration, would you hesitate to explain to someone that you had your hair colored, your nails trimmed or your legs waxed, all in the arena of fitness and grooming. Is it vanity to want to look your best? It is sometimes possible that someone's job may depend on it such as being in the military or in sales and there is nothing wrong in wanting to look and feel your best.

If you have decided to go through with Smartlipo laser surgery, after the consultation you will receive pre-op instructions to prepare for you day of surgery. Follow the instructions give—please!

Don't be alarmed if the surgeon wants you to have blood test, Urinalysis, or even a chest x-ray. The doctor wants to make certain that you are in good enough health to undergo even this rather mild form of surgery.

Considerations in Consultation Session

SmartLipo is a safe, affordable and permanent alternative. Things to consider:

- You want a procedure that is FDA approved. – The Smartlipo device was approved in November of 2006, and today is one of the fasting-growing, safest procedures in the cosmetic surgery field.

- You don't want loose skin as a result of the procedure. – Many patients of traditional liposuction find that their skin does not adjust to the reduction of fat deposits. This tendency often results in unattractive "droops."

- Patients with hanging skin or "drooping" often require skin lifts that can lead to scars, thus rendering the initial liposuction treatment useless. But by pulling the surrounding skin inward, SmartLipo laser technology can reconstruct the layers immediately under the patient's skin, completely eliminating skin drooping.

- You want something more affordable than traditional liposuction. – The cost of the SmartLipo procedure depends on how many areas are being treated, but just your neck starts at $2,000. The average traditional liposuction procedure on an abdomen costs about $6,500. The average SmartLipo abdomen procedure costs $5,000. This difference in pricing represents a 30% cost savings.

- You want permanent results. –Once the fat is removed from a particular area of the body, it will never return. If you gain fat in the future, it will be deposited elsewhere on your body.

- If you are taking time off from work to have the Smartlipo procedure, schedule more time than you may need. It the way to go in order to be safe.

Don't be alarmed if the surgeon wants you to have additional blood tests, heart tracing or electrocardiogram or even a chest x-ray. Normally, all that is required for a healthy individual is an urinanalysis and a hematology profile.

But the doctor may want to make certain that you are in excellent health to undergo even this mild form of surgery. As much as you want to get rid of it, the surgeon will likely be conservative in the amount of fat removal. More fat can always be removed later, but it is harder to replace excess fat removal.

With any surgery you need to know how to prepare. The family ought to come first and them common sense arrangements to be recovering at home for a few days. Make sure you plan you days so that the household duties and activities and go on pretty much as usual.

If you have decided to go through with having some of your fat melted and remove, then surely it is you decision. Take counsel from those you truly respect, but in the end you have to be the one determined to do something about the areas of fat in your body. Stand firm. It is your body.

Never have surgery because you are depressed. You need to have an acceptable level of positive thinking. The best way to gain this is to look to what you can become both physically and mentally.

Smartlipo is about the best quick fit you will make with cosmetic surgery, but it will only endure so long as you are willing to control food the food you put into your body.

As much as you want to get rid of it all, the surgeon will likely be conservative in the amount of fat to be melted and removed. It is always better to be conservative than not.

Smartlipo Surgical Procedure

Treatment is carried out in a sterile surgical environment. Surgery is most often performed as an outpatient procedure, but in extreme cases, or those where other medical conditions present cause for concern, an overnight hospital stay may be recommended. Many hospitals have a special "23 hour" outpatient package. The surgery itself usually takes less than an hour to complete. However, more extensive procedures may take longer.

Markings of Areas of Fat

During the "prep" phase for Smartlipo, you will be taken into the surgicenter where the laser treatment will be. You will then be marked with a sterile marker to indicate the areas to be treated and where the incisions will be carefully placed. Usually, these markings are made while you are standing. The special pencil marking lines are traced outwards from one single point in which the local anesthetic is injected.

You will also be scrubbed with an antimicrobial agent to minimize the possibility of infection. Then you will be connected to monitors that track your vital signs and an intravenous (IV) line will be started.

A few patients stand during the Smartlipo procedure, depending on the area of the body being treated. For example, it is sometimes helpful for a patient to be standing while removing fat from the hips or thighs.

Smartlipo Surgery

At the beginning of the procedure, the area to be sculpted will be filled with the tumescent fluid that helps numb the site and shrink the blood vessels. This fluid makes the area wet and full, and causes the fat cells to easily liquefy with the Smartlipo procedure. The local anesthesia is administered.

The optical fiber of the laser is inserted into the same incision created for the infiltration of the anesthetic solution.

As the procedure begins, in order to facilitate the insertion of the small cannula, a small incision must be made with a micro-scalpel or punch method. This tiny puncture is usually made at a crease in the skin and the cannula is inserted at these points. The incisions required for the procedure are so small that they require no stitches following the procedure.

As the procedure progresses, the laser is turned on and the small cannula enters the opening where the incision is made. The surgeon will then proceed to move the optical fiber into the adiposity (fat) that is to be dissolved located just under the skin. You need not be concerned that the small cannula will penetrate to the depth of vital organs and endanger them. The probe and firing of the laser is restricted to the layer of fat just under the surface of the skin where there is no danger of harm to the body during the firing of the laser. The actual treatment begins by slowly moving forward, following the lines traced on the skin and constantly maintaining the same speed of movement.

When the laser is inserted into the skin and heats the fat cells from 35 to 45 degrees centigrade, it results in a breakdown of membranes and a release of liquid fat. The laser does little harm to other tissue, including nerves, blood vessels, and connective tissues.

For the most part, the procedure is bloodless. There is a smaller amount of blood loss than that taken in the lab test. The laser seals off much of the tiny blood vessels and interacts with the dermis resulting in strengthening the collagen.

The surgeon will follow the carefully pre-marked lines to reshape the areas to be treated.

The method of breaking up the fat is performed by hand. The surgeon moves the cannula back and forth in several directions, allowing the laser beam to strike the fat tissue and melt it, causing it to become a thick liquid, much like sour cream or buttermilk.

An advantage of this procedure is that, whereas surgeons using standard liposuction methods must make straight back-and-forth movements, the tiny Smartlipo cannula can bend around curves in the body. During surgery, if there is a large area to be treated, a standard liposuction device is sometimes used first and then the Smartlipo clears away the smaller areas where fat is concentrated.

Pain Control

The actual procedure is generally well tolerated by most patients. This is due to the local anesthesia. You may feel a tugging sensation during treatment, but you should not feel any pain. You may experience minor swelling, bruising or tenderness for a few days, but because no suction is applied following Smartlipo, it is safer than other traditional methods used in liposuction.

For some areas of the body, one reason for use of minimal pain killers is to have you stand before the procedure is completed. By doing this, the surgeon can judge for symmetry and balance before concluding the operation. There is little post operative pain associated with Smartlipo. There may be some numbness, but little pain.

Time Required

The length of time required to complete the surgery depends on the technique used and the amount of fat to be removed, but it can vary from as little as 30 minutes to over an hour. Smartlipo may be done in conjunction with other cosmetic procedures. Cleaning and garmenting the patient will take longer.

Additional Treated Areas

The cannula may be reinserted into a second incision site for more fat to be removed in the targeted area. This can be repeated in several areas until the surgeon with an artistic eye determines that enough fat has been dissolved. It is possible while still performing the procedure for the surgeon to work on adjoining areas as well.

Additional Sessions

It is possible to have more than one surgical session. Whether there is a need for a second or third session depends on the area and location. Generally 1-3 sessions are more than adequate for most individuals. Repeat sessions are usually done after an interval period of 3-6 months or even a year later.

Recovery from Surgery

Smartlipo is so revolutionary that the procedure actually helps you recover faster. Because powerful laser technology is incorporated into the Smartlipo device, it is minimally invasive, which means recovery is minimally difficult.

One of the most challenging aspects of conventional liposuction recovery is getting over the anesthesia. Since local anesthesia at the procedure site is most often all that is needed, Smartlipo has none of the risks or recovery issues associated with general anesthesia.

Some of the body areas being treated will require the patient to wear a compression garment for several days after the surgery to reduce the possibility of swelling and to help ensure a sculpted Smartlipo result. The patient will change to a secondary garment for perhaps a week following surgery to insure a good result.

Patients describe mild to moderate muscle ache at the procedure site. Many describe the feeling as the same soreness level after completing an intense workout. You should be up the day following your Smartlipo procedure, but avoid strenuous physical activity for 14 days. Walking at a mild to moderate pace for the first few days after your procedure can be helpful.

You want to have someone accompany you at this stage of recovery, should you need to use the restroom. The surgeon will have you remain in the immediate area until you feel comfortable before having someone take you home. If you had thigh, abdomen, or leg treatment you will be given a compression garment to wear for several days while the swelling goes down.

Rapid Recovery

Normally you will be back to work or other activities in a short time. Your regime for rapid recovery may include vitamins, prescription medicines (if any pain) and rest. It is very important that you carefully follow the advice of your surgeon after treatment to help ensure the success of the procedure and reduce the risk of complications.

Bruising

Any bruising in the area where Smartlipo has been applied disappears usually in 2-3 days and you will be able to resume normal activities within 24 hours, but not full exercise of those areas for 10 to 14 days. The mild ache can be relieved with over-the-counter pain medications such as Tylenol or Advil.

Results of Smartlipo

Shortly after the Smartlipo treatment, the treated area will appear more compact, tighter and smaller. Within the next 6-8 weeks, the skin and surrounding area will adjust to the body contour. The process of contouring is a gradual process and the best results are seen over the following months.

Return to Work

Many patients can return to work in several days. You may wish to take off a couple days, but this is optional with most treatments performed with Smartlipo.

Fat Gain

Adults do not create new fat cells. When a person gains weight, it is caused by fat cell expansion. The Smartlipo procedure permanently removes the fat cells. Should you gain weight, most fat will be deposited in areas where the procedure was not performed.

And while Smartlipo can improve your self-confidence, it may not match your mental ideal. Nevertheless, Smartlipo is a safe, permanent alternative to removing bulges of fat.

Thrilled with their Smartlipo Results

The experience was great. You are awake for the whole thing. You can feel something but no pain that's for sure. The doctor marked all the areas ahead of time and made sure that everything was even.

Judy F.

I had Smartlipo on my neck/double chin. It is now day 3 and the results are significant. I have a fair amount of bruising, but it fades every day. I'm also slightly swollen, but that, too, is better each day. My surgeon told me to expect to see 50% of the result within the first few days, and the remaining percentage over next few months as the skin tightens. I had a little discomfort, but no pain, during the procedure. I had very minimal pain on the first day. I had a pain pill to sleep that first night, but haven't needed anything since. I'm faithfully wearing my compression garment as directed, and I think that is helping the swelling. I could have returned to work after 2 days, but I'm glad I scheduled some time off, because I want the bruising to be mostly gone before going back. I'll go back to work on day 6, and I'm hoping I'll look fairly normal by then. I have about 10 pounds to loose, then I'm going back to get my abs and love handles done. I think this procedure is terrific, but it's important to be realistic.

Jim G.

5

There are alternatives to Smartlipo, but few will deliver the results of Smartlipo.

Other Methods of Fat Removal

Traditional liposuction has been available for several decades and is the best procedure for treating large areas of the body where fat deposits prevail in mostly obese patients. Recovery with traditional liposuction is not only invasive surgery, requiring sutures, but it is typically slower healing during recovery. With traditional liposuction, bruising is more extensive, and with many overweight patients, the results lack a well-defined sculpture of the body. It may result in loose, less attractive skin that may require a body lift. Although traditional liposuction leaves less than ideal results, it is a marvelous method for removing large amounts of fat. In the past couple of years, liposuction has greatly improved since its inception in the 1980s. This is due to power-assisted (PAL) and ultrasonic-assisted methods of removal.

Power-assisted liposuction involves use of a suction tube (cannula) that vibrates at high speed to loosen deposits of fat, making them easier to resolve. Most surgeons who perform the latest liposuction procedure use a tumescent technique to prepare and desensitize the fat tissue before removal. Along with the cannula's high-speed vibration, it has become a safer procedure, making a quicker recovery as compared to traditional liposuction.

With Power-assisted liposuction for large fat deposits, too large for the Smartlipo procedure, the combination offers the best method for obese patients.

In addition to PAL, surgeons perform ultrasonic-assisted liposuction. With this method, ultrasonic sound waves are directed into the treatment area, rupturing fat cells and allowing for easier removal. It is often used for fibrous areas of fat. There is also suction-assisted liposuction (SAL) available with many surgeons.

Any one of the above may be combined with Smartlipo to gain the best possible results, but none compare to the results of Smartlipo. See the chart displayed here to make you own judgment.

The Advantages of Smartlipo

- A Much Smaller Incision — half a grain of rice incision in skin with less trauma than liposuction.
- Less Bleeding and Bruising – Destroys fat cell by melting the fat with less trauma to the body.
- Rapid Recovery — little pain, if any; pressure garments less common; lunch break procedure.

Treatment Chart for Smartlipo

Before Treatment		
	Smartlipo	**Liposuction**
Treatment By	Surgeon	Surgeon
Consultation	Free	
Treatment on Same Day	Yes. Possible	Usually not
Treatment		
Treatment Time	1 hour	1-2 hours
Surgical Invasiveness	Minimal	More invasive
Needle	1 mm fiber optic laser	3-6 mm cannula
Anesthetic	Local	General possible
Fat Cells	Melted and absorbed	Vacuumed
Blood Loss	Minimal	some blood loss
Obesity Treatment	Localized fat deposits	Yes
Area		
Face	Yes	Yes
Forearm	Yes	yes
Knees	Yes	yes
Inner thighs	Yes	yes
Upper abdomen	Yes	yes
Repeatable	Yes	yes
After Treatment		
Recovery time	0-1 day	3-10 days
Pain Level	mild	Often felt
Compression bandage	2-7 days	6 weeks+
Bruising	very little	Medium-High
Swelling	very little	Medium-High

Results fully visible	3 – 4 months	3 months + sagging skin
Skin tone	Tightened by Laser	Sagging possible
Permanent fat removal	Yes	Yes
Scarring	Pinhole	Small-Large
Infection	minimal	Possible

6

The New MPX Laser Instrument is the next generation for fat removal.

The next generation Smartlipo is the new MPX mutiplex that delivers power and speed. The energy is delivered safely and responsibly and supported by extensive clinical studies and investigation. Here are some of the advantages of this next generation Smartlipo instrument:

- Enhances hemostasis for less trauma.
- Disrupts a broader region of adipose tissue.
- Heats treated areas evenly.
- Enhances skin tightening by heating collagen bundles in dermis.

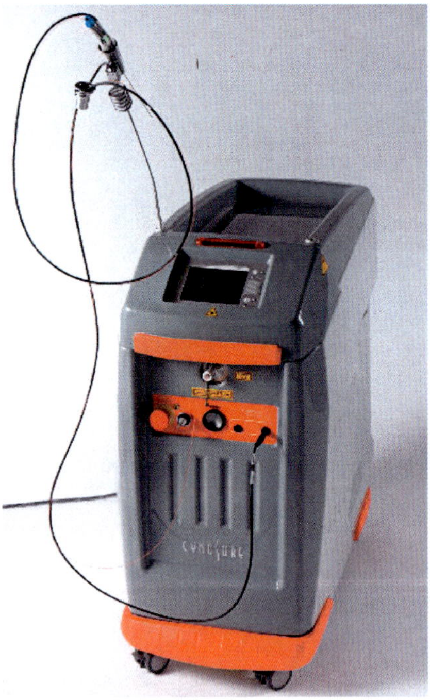

Smartlipo MPX represents Cynosure's next generation of Laser Body Sculpting Technology

This latest instrument is an intelligent delivery system that allows a precise distribution of energy and safety and consistent patient outcomes. This new system has these advantages:

- Detects movement within the handpiece.
- Calculates laser energy to provide even distribution of energy according to the speed of movement.
- Intelligence within the machine determines the appropriate energy based on Sensitivity Rating.

According to one of the leading surgeons in the use of the new MXP machine, Dr. Bruce Katz, "Smartlipo MPX Laser is much faster and more powerful than the last generation of Smartlipo. The dual wavelength technology precisely targets larger areas on the body, causes less bleeding, swelling and bruising, and dramatically tightens skin around the treated area."

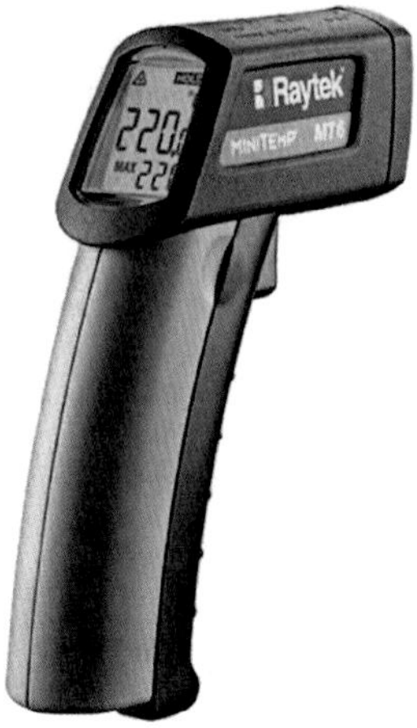

This handheld gage monitors skin temperature with external thermal probe

The innovative MultiPlex technology is the only dual-wavelength laser system that has been cleared by the Food and Drug Administration (FDA). This innovation in Smartlipo laser causes fat liquefaction and skin tightening.

7

There are questions to be answered.

Questions and Answers

How many treatments will I need?
For most patients, typically only one treatment is required. The session may last from 45 minutes to one hour for each area being treated.

Do I have to do anything post-treatment?
After the treatment, you will have minimal pain or bruising. We recommend 1-2 days rest but some patients are able to return to work immediately (the so-called Lunch hour lipo).

Can Smartlipo treat cellulite?
The procedure will improve the body's shape and may improve cellulite to a certain degree, tightening saggy skin.

Does Smartlipo produce lasting results?
Yes. Fat cells do not reproduce in the body. Once they are removed, there is only the problem of expanding the remaining cells due to food intake.

Does Smartlipo respond well to all skin types?
Yes! Smartlipo works on people with light or darker skin. The laser is below the skin targeting the fat cells and the underside of the dermis. The melanocytes are NOT affected.

Many men and African-Americans have a tougher and more fibrosis fat that may be more difficult to shape/sculpt with standard liposuction techniques. The use of Smartlipo minimizes any difference.

Can more than one area be treated in a single session?
Yes. You may treat up to six areas, depending on the extent of correction. Speak to your cosmetic surgeon to find out more information.

What happens immediately following the procedure?

A special garment may be required to facilitate healing and compress the area. There may be some swelling and discomfort, but any pain is typically considered simply a dull ache. The rate of bruising, if any, is mild.

How long do I have to wait to go back to work?

If you don't need to exert too much physical effort at your job, you can go back the same or next day. For jobs that are more strenuous, consult your surgeon for advice. Where there has been fat removal over larger areas you may be able to return to work after a weekend.

When do I see results?

The patient will look significantly smaller the day after surgery. Over a course of 6 weeks following a procedure, you'll be able to see dramatic improvements as the swelling diminishes. You can expect to see the final results in approximately 2-4 months. However, this depends on the extent of treatment and your health. Follow your surgeon's instructions as closely as possible to ensure a quicker recovery.

Is it possible to repeat Smartlipo the same area of the body?

Yes, but patients must wait at least 3 months.

What are the risks and side effects of Smartlipo?

Like any procedure, Smartlipo also has its risks and side effects. These include bruising and/or discomfort, and mild pain.

What does the Smartlipo feel like?

The laser has a pleasantly warm feeling. The skin will be numbed, causing you to sense some mild heat similar to the feeling of strong sunlight.

By adding Smartlipo with traditional liposuction, is the recovery different?

Smartlipo recovery is much more rapid than traditional liposuction due to its much less invasive methods. Many patients return to work the next day. The combination will require a longer time to recuperate

What is the safety factor of Smartlipo?

Smartlipo is similar in safety to other laser treatments such as Thermage. With a skilled surgeon there is no risk of lymphatic problems related to the liquefied fat. In addition there is no risk of damaging tissue.

What is the cost of Smartlipo?

The cost of Smartlipo ranges from the very smallest removal fat for a slight double chin to a more extensive abdominal fat removal. The cost can range anywhere from $2000 to $10,000. The cost depends on the areas of the body needing fat removal. You may wish to have several areas of fat pockets removed. For each area of the body there is usually a set fee. This can be discussed at the time of consultation.

What complications may occur?

Surgical complications of Smartlipo are quite rare. They are less than 1 percent. Like any surgical procedure, Smartlipo is associated with common side effects such as slight bruising, swelling, and numbness. Skin irregularities have been greatly reduced due to the very small cannulas that result in a more even and less traumatic reduction of fat.

Can fat re-accumulate in the contoured areas?

Fat cells that are permanently removed are thought not to multiply after puberty. As long as you do not gain excessive weight, the new silhouette is permanent. Fat removal however, does not allow for abuse of diet later.

Patients, after Smartlipo, have sometimes found that areas treated by the laser retain the new contour whereas areas not treated have been new sites for fat development.

If you gain weight, the size of the individual fat cells will increase, and too much weight gain may mask the benefit of the procedure. The benefit will show when the extra weight is reduced. People with prior weight problems still have to watch their weight and exercise.

How are the benefits of Smartlipo measured?

The cosmetic success of Smartlipo relies on appearance and not scales. Though little fat may actually be removed, dramatic change in silhouette is often possible. The most important thing is your satisfaction and increased self-acceptance of the change in appearance.

Is Smartlipo right for me?

The best way to decide is to speak with a surgeon who has experience and specializes in surgical fat removal and in facial and body sculpturing. It is important that the surgeon has sensibility and an artistic touch.

How will my skin look after surgery?

Stretch marks, dimples, and ripples may remain less than a 100% improved, Even so, a skilled surgeon can smooth out much of these distressing issues of body fat. In some cases, cellulite may appear visibly reduced. The changes following Smartlipo are analogous to those who expect losing an equal amount of fat in localized areas by dieting alone. Smaller cannulas allow the fibrous connections between the skin and body to remain. These contract or shorten during healing.

Where will the procedure be performed?

Smartlipo is usually performed in a specially equipped in-office surgical suite. Smartlipo can be performed completely by local anesthetic. Single areas, i.e., neck, abdomen, knees, outer thighs, etc., and multiple areas can be treated in a given session, depending on the number and size of the areas.

You will be given specific instructions on how to prepare for surgery, including guidelines on eating, drinking, and taking certain vitamins and medications. Smokers should plan to stop smoking for a minimum of one to two weeks before surgery and during recovery. Smoking decreases circulation and interferes with proper healing resolution of swelling.

Glossary

Abdomen: The belly, that part of the body that contains all of the structures between the chest and the pelvis. The abdomen is separated anatomically from the chest by the diaphragm, the powerful muscle spanning the body cavity below the lung.

Abnormal: Not normal. Deviating from the usual structure, position, condition, or behavior. In referring to a growth, abnormal may mean that it is cancerous or premalignant (likely to become cancer).

Anesthesia: Loss of feeling or awareness. A general anesthetic puts the person to sleep. A local anesthetic causes loss of feeling in a part of the body such as a tooth or an area of skin without affecting consciousness. Regional anesthesia numbs a larger part of the body such as a leg or arm, also without affecting consciousness. The term "conduction anesthesia" encompasses both local and regional anesthetic techniques. Many surgical procedures can be done with conduction anesthesia without significant pain. In many situations, such as a C-section. Conduction anesthesia is safer and therefore preferable to general anesthesia. However, there are also many a type of surgery in which general anesthesia is clearly appropriate.

Cannula: Simply a slender tube that can be inserted into a body cavity or duct. During the insertion of the cannula,

Cellulite: Popular term for deposits of fat that have a cottage cheese-like or puckered texture. Medically, cellulite is not considered abnormal.

Chest: The area of the body located between the neck and the abdomen. The chest contains the lungs, the heart and part aorta. The walls of the chest are supported by the dorsal vertebrae the ribs, and the sternum.

Dizziness: Painless head discomfort with many possible causes including disturbances of vision, the brain, balance (vestibular) system of the inner ear, and gastrointestinal system. Dizziness is a medically indistinct term which laypersons use to describe a variety of conditions ranging from lightheadedness, unsteadiness to

Fat: 1 Along with proteins and carbohydrates, one of the three nutrients used as energy sources by the body. The energy produced by fats is 9 calories per gram. Proteins and carbohydrates each provide 4 calories per gram. **2** Total fat; the sum of saturated, monounsaturated and polyunsaturated fats. Intake of monounsaturated and polyunsaturated fats can help reduce blood cholesterol when substituted for saturated fats in the diet. **3** A slang term for obese or adipose. **4** In chemistry, a compound formed from chemicals called fatty acids. These fats are greasy, solid materials found in animal tissues and in some plants. Fats are the major component of the flabby material of a body.

Hematoma: An abnormal localized collection of blood in which the blood is usually clotted or partially clotted and is usually situated within an organ or a soft tissue space, such as within a muscle.

Laser: A beam of light that can produce intense heat when focused at close range. Lasers are used in medicine in microsurgery, cauterization, for diagnostic purposes, etc. For example, lasers are employed in microsurgery to cut tissue and remove tissue.

Liposuction: The most common cosmetic operation in the U. S. with over 400,000 such surgical operations done annually, liposuction involves the surgical suctioning of fat deposits from specific parts of the body, the most common being the abdomen (the "tummy"), buttocks ("behind"), hips, thighs and knees, chin, upper arms, back, and calves.

Nerve: A bundle of fibers that uses chemical and electrical signals to transmit sensory and motor information from one body part to another.

Postoperative: After surgery. As opposed to preoperative, before surgery. Before surgery. As opposed to postoperative, after surgery.

Pulmonary: Having to do with the lungs. (The word comes from the Latin pulmo for lung).

Pulmonary embolus: A blood clot that has passed into and occluded the lung's pulmonary artery. An embolus causes an embolism. In this case, the embolus, a clot or foreign material, has been carried through the blood into the pulmonary artery or one of its branches, plugging the vessel.

Relaxant: Something that relaxes, relieves, reduces tension. For example, a muscle relaxant is often administered during abdominal surgery to relax the diaphragm and keep it from moving during the surgery. Repose, relaxation

Sedative: A drug that calms a patient down, easing agitation and permitting sleep. Sedatives generally work by modulating signals within the central nervous system. These sedatives can dangerously depress important signals needed to maintain heart and lung function if they are misused or accidentally combined, as in the case of combining prescription sedatives with alcohol. Most sedatives also have addictive potential. For these reasons, sedatives should be used under supervision, and only as needed.

Skin: Skin is the body's outer covering. It protects us against heat and light, injury, and infection. It regulates body temperature and stores water, fat, and vitamin D. weighing about 6 pounds, the skin is the body's largest organ. It is made up of two main layers; the outer epidermis and the inner dermis.

Smartlipo: A laser assisted method of removing fat by melting it.

Suture: 1. a type of joint between the bones of the skull where the bones are held tightly together by fibrous tissue. **2.** Thread-like material used to sew tissue together. **3.** To stitch a wound closed. The word suture came with little change from the Latin *suture*, "a sewn seam." In Latin, the verb *suere* is "to sew, stitch, or tack together."

Thyroid: 1. the thyroid gland. Also, pertaining to the thyroid gland. **2.** A preparation of the thyroid gland used to treat hypothyroidism. **3.** Shaped like a shield. (The thyroid gland was so-named by Thomas Wharton in 1656 because it was shaped like an ancient Greek shield.)

Tumescent: Swelling; slightly tumid (swollen).

Tumescent liposuction: The surgical suctioning of fat deposits from specific parts of the body, the most common being the abdomen (the "tummy"), buttocks ("behind"), hips, thighs and knees, chin, upper arms, back, and calves after pumping several quarts/liters of a salt water solution underneath the skin of the area to be suctioned.

Photos of Sculpted Patients

The following are photos for you to view some of the hundreds of fat reduction patients that I have performed over the past two years.

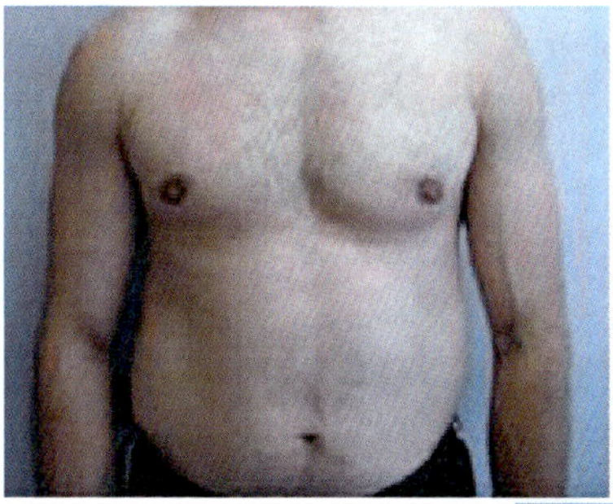

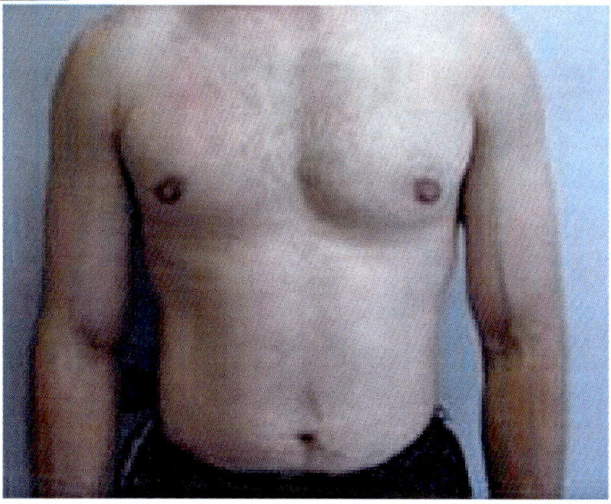

Patient in his 30's received Micro-Lipo of the upper abdomen, lower abdomen & waist

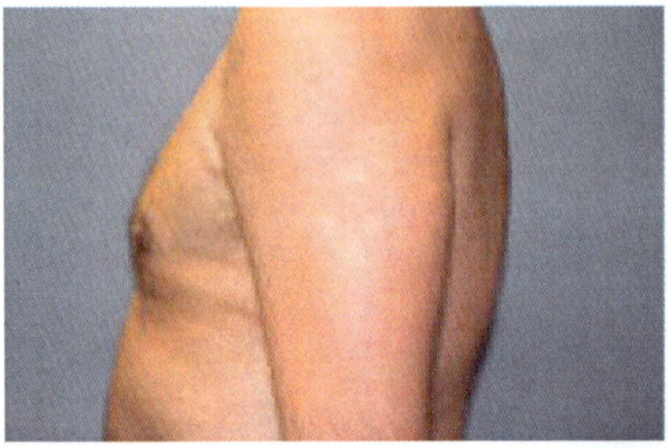

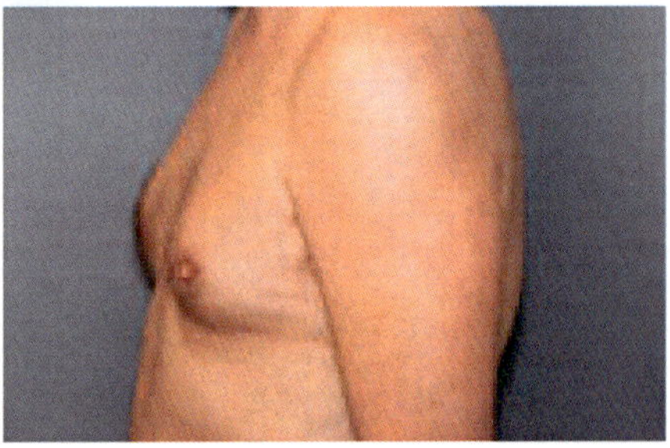

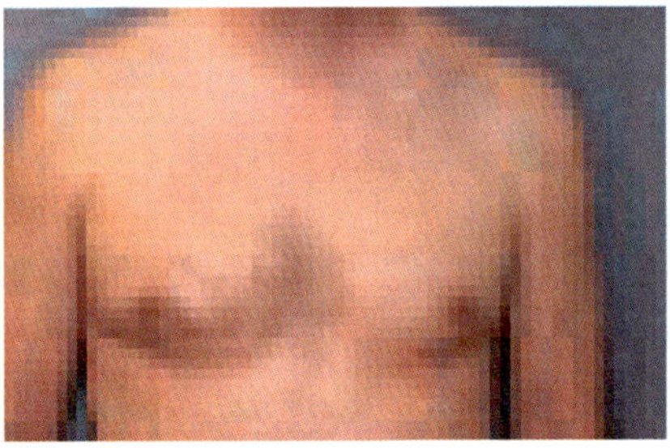

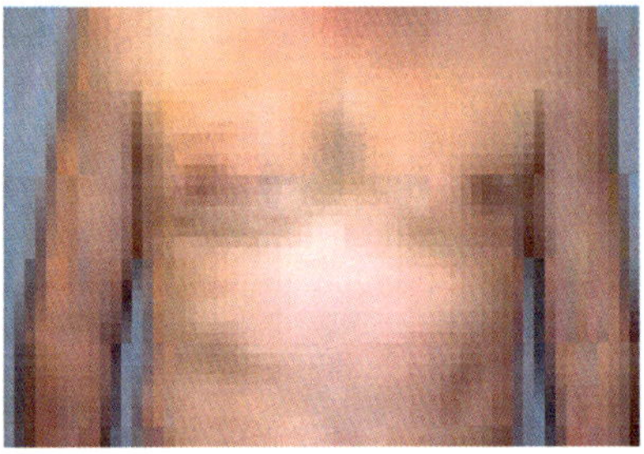

40 year old man who has struggled all his adult years with a larger chest than he would like. Dr. Schafer performed the gynecomastia or male breast reduction procedure with the assistance of the newest technology called SmartLipo.Post operative photos are 3 days later.

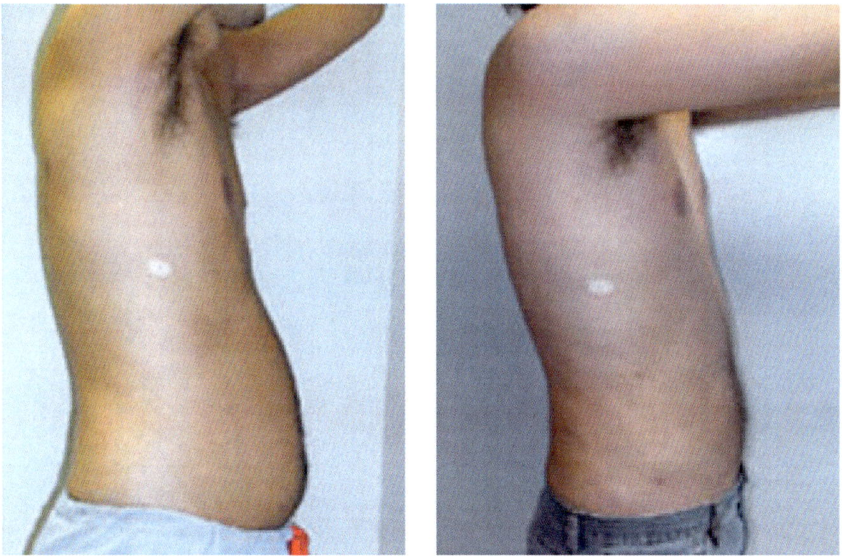

Patient in his 30's received sculpting of the upper abdomen and lateral waist.

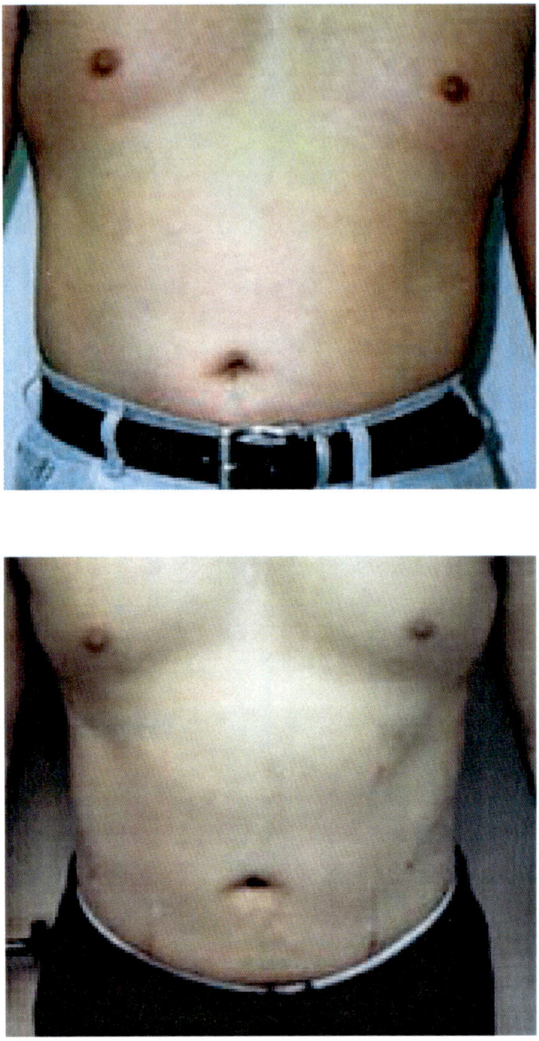

Patient in his 40's received sculpting of the upper abdomen and lateral waist

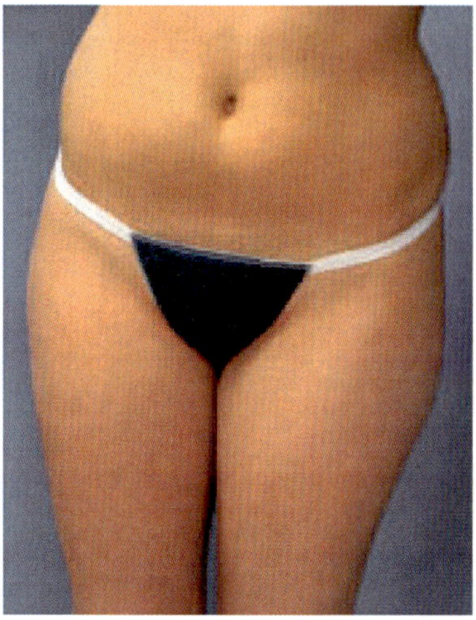

Before

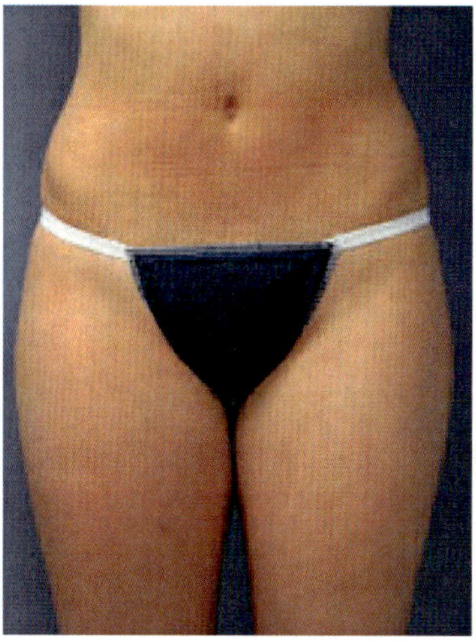

After

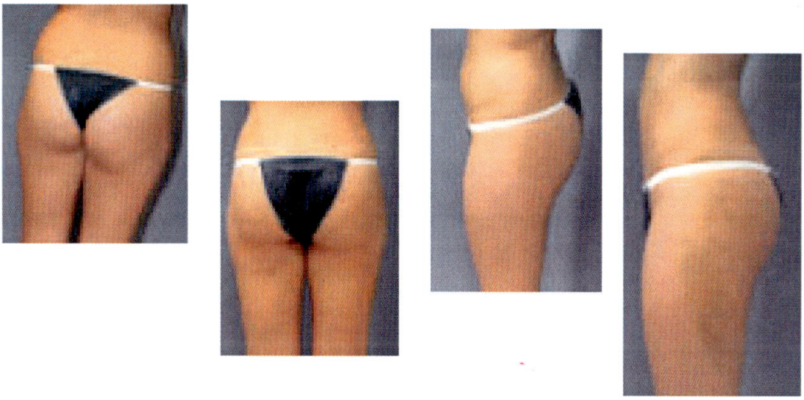

7 days post op Smartlipo of the upper and lower abdomen,
outer thighs and hips

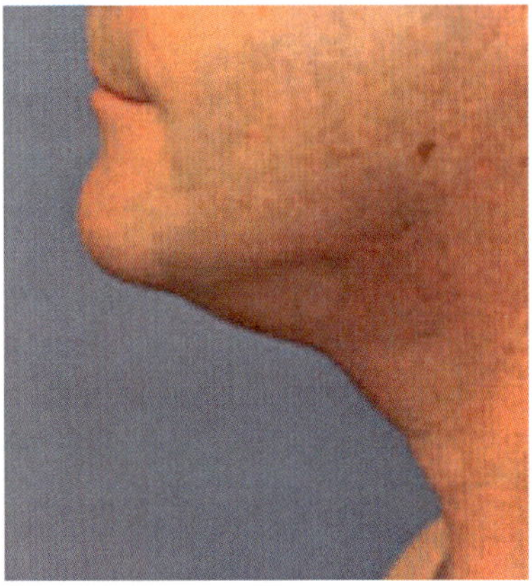

Before

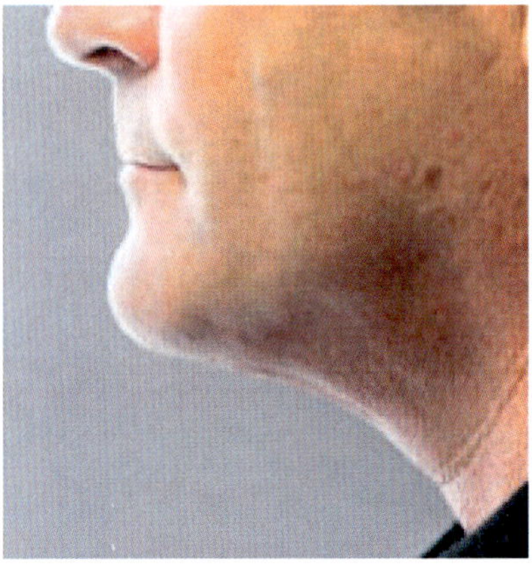

After

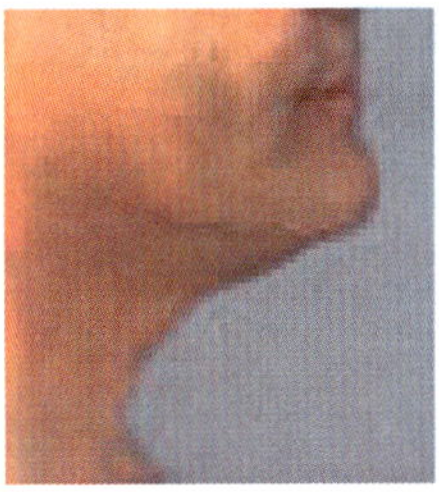

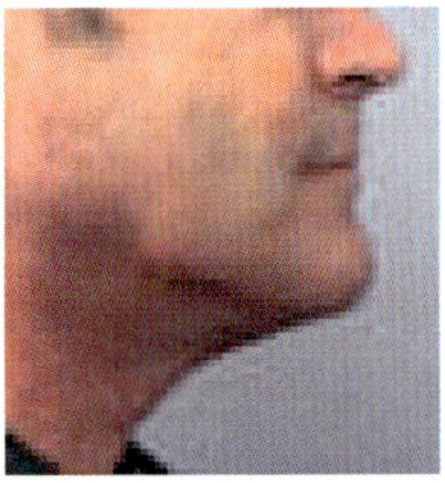

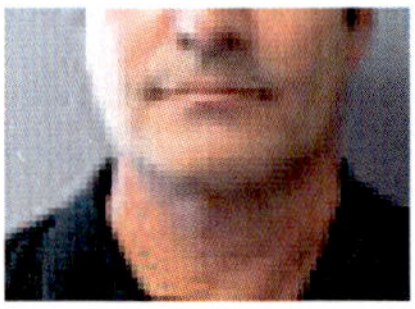

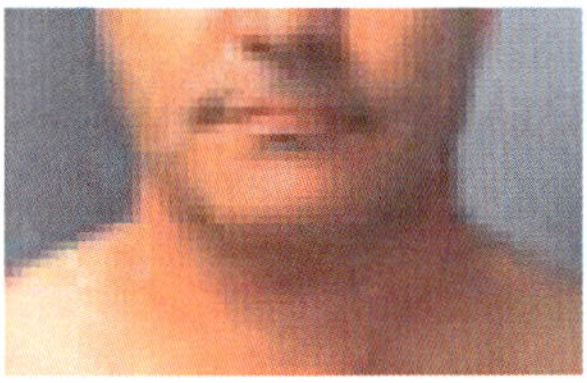

53 yo male had SmartLipo of the Neck, Chin and Jowls.
Photos are 2 months post op.

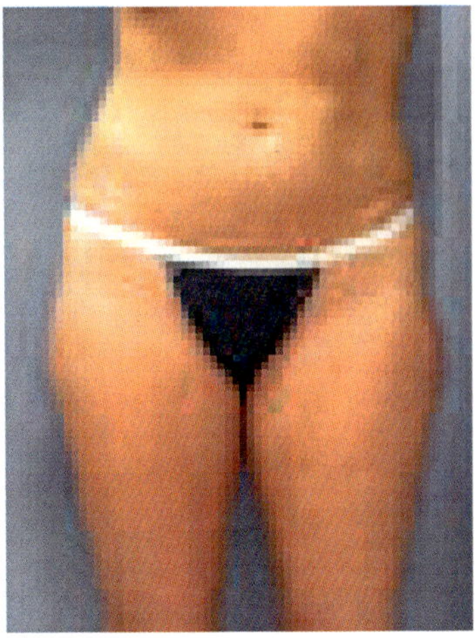

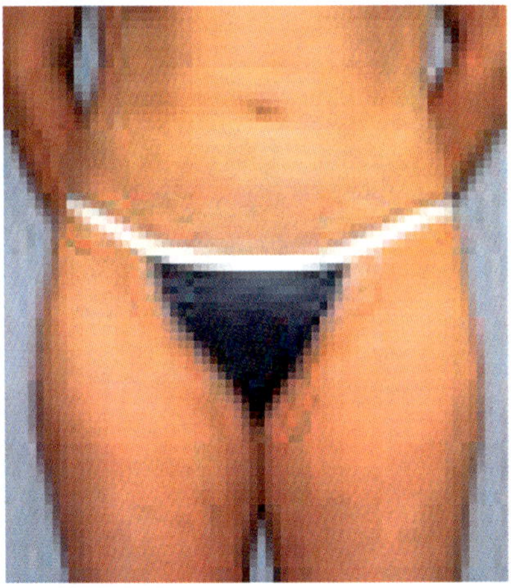

38 year old female who flew into San Diego for Smartlipo with Dr. Schafer. Patient desired thinner legs and a flatter stomach. Mission accomplished with Smart lipo of the upper and lower abdomen, lateral waist, hips, inner and outer thighs. Photos taken 1 week after surgery.

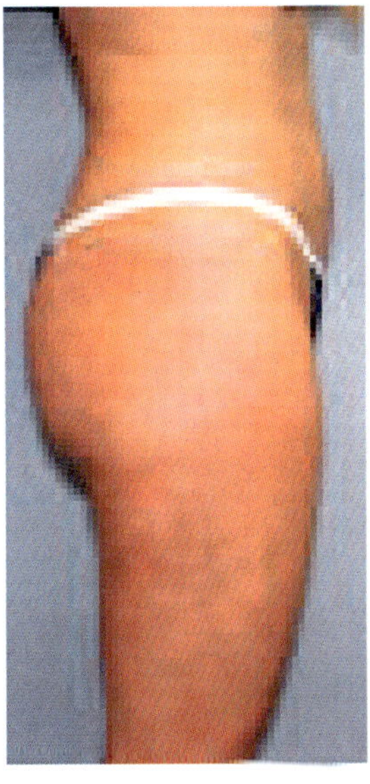

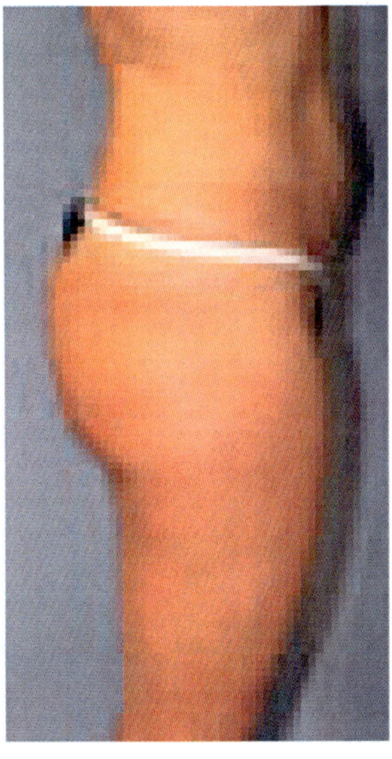

About the Author

Jeffry B. Schafer, MD is a fully trained Cosmetic & Reconstructive Surgeon and has been in practice since 1978. Along with his 25 years of experience in cosmetic surgery, he has 10 years of lipo sculpture experience. Dr. Schafer was a former staff physician with Scripps Clinic and Research Institute in La Jolla. He was a clinical instructor at The University of Alabama School of Medicine. Dr. Schafer also taught at the University of California at San Diego.

He has been an integral part of the San Diego community. Dr. Schafer has been a board member of Mainly Mozart, The San Diego Opera and Partners Against Crime. Dr. Schafer feels strongly about giving back to the community and is a supporter of the Salvation Army, numerous children's charities and has been involved in the Spirit of Caring Mobile Healthcare Clinic in Chula Vista. He has also volunteered his plastic and reconstructive skills for the Rotary Foundations Thousand Smiles in Mexico. Dr. Schafer was awarded fellowships at King's College of The University of London, England, and The Armed Forces Institute of Pathology at Walter Reed Hospital, Washington, DC.

Today he is one of the leading Smartlipo surgeons in the United States.

Made in United States
North Haven, CT
06 November 2026

21797419R00062